D1070909

THE BIOLOGICAL REVOLUTION

Applications of Cell Biology to Public Welfare

THE BIOLOGICAL REVOLUTION

Applications of Cell Biology to Public Welfare

EDITED BY
GERALD WEISSMANN, M.D.
New York University School of Medicine
New York, New York

PLENUM PRESS · NEW YORK AND LONDON

Library of Congress Cataloging in Publication Data

Main entry under title:

The Biological revolution.

Includes papers from a symposium held at the first International Congress on Cell Biology, Boston, 1976.
Includes index.
1. Cell research–Congresses. 2. Medical research–Congresses. I. Weissmann, Gerald. II. International Congress on Cell Biology, 1st, Boston, 1976. [DNLM: 1. Cytology–Congresses. 2. Public health–Congresses. W3 IN345E 1st 1976b QH573 I615 1976b]
QH583.B56 574.8′7 79-12369
ISBN 0-306-40241-6

A Division of Plenum Publishing Corporation
227 West 17th Street, New York, N.Y. 10011

Printed in the United States of America

Contributors

BRUCE N. AMES, Department of Biochemistry, University of California, Berkeley, California 94720

DON W. FAWCETT, Hersey Professor of Anatomy, Harvard University Medical School, Cambridge, Massachusetts 02138

EMIL FREI III, Sidney Farber Cancer Institute, Boston, Massachusetts 02115

ERIC R. KANDEL, Division of Neurobiology and Behavior, Departments of Physiology and Psychiatry, College of Physicians and Surgeons, Columbia University, New York, New York 10032

EDWARD M. KENNEDY, U.S. Senator (D-Mass.), Senate Office Building, Washington, D.C.

LEWIS THOMAS, President, Memorial Sloan–Kettering Cancer Center, New York, New York 10021

GERALD WEISSMANN, Professor of Medicine and Director, Division of Rheumatology, New York University School of Medicine, New York, New York 10016

CARROLL M. WILLIAMS, Benjamin Bussey Professor of Biology, Harvard University, Cambridge, Massachusetts 02138

Contents

Preface

Basic biological research is not in trouble, but support for this cultural product is perhaps more fragile than it should be. We have developed, in this country, in Europe, and in Japan a triumphant record of research accomplishment that has revolutionized our vision of the cell, the body, and the environment. The genetic code has been unraveled, the means of neuromuscular transmission have been elucidated, we know the fine detail of the cell's small geography, and we can describe the genes of mouse and man in chemical terms. But it is less clearly perceived by laity and scientist how this new revolution in biology has been of use in the service of a better medicine or environment.

The aim of this series of essays, which grew out of a symposium at the First International Congress of Cell Biology, is to instruct us all in the *use* that has been made of our new knowledge. In the fields of cancer, behavior, reproduction, genetic engineering, and environmental monitoring, practical results of our new knowledge are already apparent and more are just at hand. But neither the general public nor scientists working in their own narrow disciplines necessarily appreciate these developments. Perhaps it is all just too new, too "experimental," to permit anyone to gain a proper perspective on what we *have* done or have the potential of doing.

It is often said that the gap between research and its application is wide, that the "gap" constitutes a kind of chronic crisis. Examination of these essays should relieve these apprehensions. Basic research is probably as rapidly applied as considerations of safety permit, and in our kind of society the open forums of meetings and journals, and the structure of our research-based universities provide the means whereby applications flow from discovery. But this flow is usually within the boundaries set by disciplines: the entomologist rarely has the chance to talk with the cancer chemotherapist, the neurobiologist with the cell biologist working on reproduction. This series of essays marks one such effort of scientists from a number of special disciplines to tell each other, and us, a story of how their craft has been of general interest and general benefit to the public welfare.

This volume represents an attempt by the Public Policy Committee of the American Society for Cell Biology to tell the story of how basic research can help promote the public welfare.

Gerald Weissmann

New York

Biomedical Science in an Expectant Society

A Speech Delivered to the International Cell Biology Congress

EDWARD M. KENNEDY

Dr. Thomas, distinguished guests, ladies and gentlemen: I appreciate the opportunity to address this First International Congress on Cell Biology.

In particular, I would like to underscore the international aspect of the congress. For cell biology—like all fields of science—can only make progress through full international cooperation that knows no national boundaries. I understand that a quarter of your participants come from abroad and I would like to add my personal welcome to these distinguished visitors.

I am especially pleased to participate in this historic scientific congress. For over the past quarter of a century, no field of

EDWARD M. KENNEDY • U.S. Senator (D-Mass.), Senate Office Building, Washington, D.C.

science has made more spectacular strides than cell biology. And over the balance of the century, no other field of science offers greater promise to increase our understanding of life and our ability to deliver quality health care.

For scientific understanding and quality health care are woven together like the strands of the double helix. Without the pursuit of basic science for its own sake, we can never achieve the creative insights that advance the frontiers of knowledge. But without a concerted effort to apply that knowledge to social goals, we cannot make the benefits of science available to all people throughout the world.

In my own work in the Senate, I find it helpful to distinguish three approaches that must be followed in parallel. First, we must vigorously support basic research. This is society's investment in the future. Each dollar we skimp on basic research today shortchanges future generations. Second, we must actively seek out and support ways to apply scientific knowledge to the solution of practical problems—such as eradicating disease, improving nutrition, and enhancing therapy. Third, we must establish health care systems that design and deliver the benefits of these applications to all our citizens equitably. This is the threefold challenge that faces us in the Congress and in the society at large. If any one of these three requirements is neglected, all will suffer.

Without basic knowledge, we cannot advance. Without specific applications, we cannot target particular diseases and health problems. And without adequate delivery systems, society is shortchanged on its investment in research—and there will be increasing pressure to curtail that investment.

Unfortunately, not all members of the Congress, of the health community, or of the public understand the need for all three of these approaches to be pursued simultaneously. Yet it is absolutely imperative for the health of our scientific enterprise, as well as for the health of our citizens, that we build a wide base of understanding of the need for these three interrelated approaches. And to build that understanding, we in the Congress need the cooperation of the research community and the help of the medical professionals and concerned citizens.

The problem is compounded by the fact that the scientist and the citizen often do not speak the same language, or at the least do not engage in a useful exchange of views. Sometimes the scientists speak down to the layman in technical language that the latter cannot understand. It's tremendously important to strive to explain the implications of scientific work clearly. We can hope that meetings like this may serve to facilitate this type of exchange.

The dialogue is also advanced by clear reports like the one presented to the President's Biomedical Research Panel by the subpanel chaired by Dr. Lewis Thomas. Dr. Thomas' report summarized very well what is happening in biomedical research and what the payoffs of this research can be in terms of basic knowledge and practical application.

In advancing this dialogue further, there are two challenges which must be met. First, the average citizen has no way of evaluating most of what goes on in science. What he does know is that he is investing well over $2 billion a year in biomedical research. And there is little doubt that what the taxpayer expects from this investment is the ultimate conquest of disease—specifically those diseases whose names he knows and whose consequences he fears.

This may represent a simplistic view of the process whereby we translate research into applications, but it is the prevailing view in the country and in the Congress. And it is the view with which we in public policy positions have to contend.

To deal with this view, we need your help. The biomedical research scientist has not told his story well to the public. He has not adequately fostered the kind of educational process which would inform the public as to the nature of science and its mode of operation. He has not made clear to the public the great contributions of biomedical research to improving public health. That is where the payoff is for the layman and that is what will assure continued support for research.

The second public challenge facing the research community is not only to educate but to demonstrate its cooperation in helping solve society's problems in the important area between the laboratory and the clinic. The challenge is there for biomedical scientists, whenever feasible, to help translate their basic knowledge into useful applications. This will work wonders in building continued public support for research.

There are a number of present and future problems to which the biomedical research community is in an especially good position to provide answers, or at least to contribute productively to a clarification of some of the issues.

The first problem is how we assure that our medical research capacity best supports the objective of improved health care and the conquest of disease. How do we assure ourselves that the fruits of the biomedical research endeavor are finding their way into clinical practice, in both a timely and economic manner? For instance, does the obligation of the individual scientist end when he publishes his research results in the development of a new diagnostic procedure or therapeutic mode?

For example, whose responsibility is it to assess the efficacy of these innovations, and to assure that both the public and the professions are being educated in their proper use? After the research develops an efficacious but expensive halfway technology such as kidney dialysis, how does society allocate this within its finite resources? How do we monitor and assess such things as heart transplantation or the extended use of CAT scanners? Or, as was recently brought out in hearings before the Senate Health Subcommittee, how should we respond to the fact that years after the demonstration that simple mastectomy plus chemotherapy for breast cancer is just as effective as the terribly disfiguring Halsted-type radical mastectomy, still over 60% of American women with breast cancer are being subjected to the radical procedure?

Is there no way that we can, as a society, improve our mechanisms for assessing health technologies? Is there no way that we can do a better job of communicating research results to both the public and the health professionals and provide them with better tools? I appreciate that these questions are ones that the research community is not accustomed to being asked. But they do reflect the deep concerns of society in the health area, and you members of the research community represent some of our nation's wisest professionals. Even when you do not have the answers, you can help the rest of us in better formulating the questions, and assist in our search for the answers.

Another major problem which the research community must help society to face is the growing number of ethical and moral issues emerging from biomedical research itself. For example, we recently confronted the perplexing problem of how to deal responsibly with the issue of fetal research. The members of the health community appreciate the importance of studies in development and differentiation in cell biology. Not only will basic knowledge be obtained, but also there are important implications for improving our understanding of the diagnosis, therapy, and prevention of certain congenital diseases. Nevertheless, there was enormous negative sentiment expressed in the House of Representatives, which voted to ban fetal research across the board.

I attempted to help cope with this problem by introducing legislation to create a National Commission for the Protection of Human Subjects of Biomedical and Behavioral Research.

This commission was established with distinguished membership from the medical research community as well as representatives of the public interest from such fields as law, ethics, and religion. I believe that the guidelines formulated by the commission to govern fetal research in this country would have failed without significant input from both the research community and the representatives of the wider public.

Another, more current example of the ethical issues and the need for public involvement is the controversy over recombinant DNA experimentation. Such manipulations of genes are certainly one of the prime examples of the biological revolution at work. We are told that the techniques have the potential for remarkable benefit to mankind in such areas as genetic engineering to cure disease, for the bioproduction of scarce substances such as hormones, or in agriculture to help advance the "Green Revolution."

On the other hand, others warn us that research in this area may lead to the formation and release of some sort of deadly "Andromeda Strain" of pathogenic organism with which the world may not be prepared to cope. Whose predictions is the public to believe in assessing the possible benefits or the possible hazards of this type of research? It was this issue which led the Cambridge City Council on July 9 [1976] to become the first municipality in the nation to assume a role in the regulation of scientific research. This is the type of question in which science and society desperately need to cooperate and make joint judgments on for our common good.

It is to the credit of the scientific community that they voluntarily met at Asilomar to declare a temporary moratorium on research in this area, until some of the problems could be examined more closely. And the hearings and careful review done by the National Institutes of Health on guidelines for the recombinant DNA research are to the further credit of the scientific community. These are excellent precedents from which both science and the public can learn for the future.

But one lesson is clear. The public must participate in the resolution of issues in which science impacts on society. They must be in on the takeoff as well as the landing. They must help formulate the key public policy questions which must be answered, and they must participate in the commissions and other groups convened to resolve these issues.

That is not to say that the public can be expected to fathom the full complexities of scientific and technical matters. But they can be expected—and are indeed entitled—to understand the impacts of science and technology on human values and social goals. In achieving this understanding they need the full cooperation of the scientific community.

This is no easy task. But with the goodwill and the aid of scientists like yourselves, I am confident it can be accomplished. One thing is certain: we have to succeed at bringing the public into such decisions if democracy is to survive in an age of science.

Of course, the NIH guidelines do not assure complete responsibility in recombinant DNA work and in synthetic gene production. These guidelines only apply at the moment to NIH research, although I am pleased to hear that the NSF has voluntarily adopted the guidelines also. What of other federally sponsored research that does not come under the control of NIH? What of nongovernmental research in this area, such as may be in progress in the pharmaceutical and other industries? This must be of concern not only to responsible citizens and their representatives in the Congress, but also to the scientific community at large.

It was because of this concern that Senator Javits, the ranking minority member of our Senate Health Subcommittee, and I sent a letter to President Ford pointing out the limitations of the NIH guidelines and recommending their extended application beyond the confines of NIH-sponsored research. I mention this because it is a good example of how science and the public can work out their mutual interests—and how the Congress can intervene only to assure that all legitimate interests are weighed to the good of our society as a whole.

Hopefully, the scientific community will continue to work closely with both the public and their congressional representatives on this and other important issues, and help refine our tools for public participation with science in such decisions. In the long run, these activities are as important to the scientist as they are to the citizen.

During the coming year, the Senate Health Subcommittee will hold a series of hearings to explore these and other fundamental issues in biomedical research. I look forward to hearing from you and other members of the scientific community in making these hearings a success. I have little doubt that the Congress will reaffirm our society's belief in basic research by the legislation we pass in the next year. But I hope we can also express in the legislation an additional commitment to apply both scientific knowledge and tools to help assure quality health care for all.

Ultimately, the scientific community, the public, and their representatives in the Congress share the same goal: We want lives as free of disease and disability as science can make them—lives enhanced by the best possible health care, which is health care supported by the best possible science. And we want these lives for all our people, not just a fortunate few. We want a society that assures the benefit of our science and our health care to all, and excludes no one because he can't pay the bill or lives in the wrong part of the city. Finally, we want a society that has the wisdom to invest its best minds and its scarcest resources in the things that do the most to enhance the dignity of our human condition. Medical science and health care are among the best investments we know how to make.

I believe we share these values. They are the values that have supported my work in health care. And they are the values that must undergird our continuing efforts to improve our nation's research efforts in the years ahead.

Thank you.

How Basic Knowledge of the Cell Cycle Has Helped Us Treat Cancer Patients

EMIL FREI III

While empiricism has, in the past, prominently influenced therapeutic research generally and therapeutic research in cancer specifically, knowledge derived from basic and bridging sciences is increasingly impacting on such research. For cancer, such sciences include particularly pharmacology, immunology, and cytokinetics. The purpose of this presentation is to describe the impact of cytokinetics on therapeutic research in cancer. Increasing knowledge of the cytokinetic behavior of tumors and target normal tissues, and of the effect of various chemotherapeutic agents on aspects of the cell cycle have provided increasingly important therapeutic leads for extrapolation to the clinic. With advances in methodology, it has become increasingly possible to conduct selected cytokinetic studies in parallel with clinical trials.

The definitions and dynamics of the mitotic cycle of the cell are presented in Figure 1. All mitotically active normal and

EMIL FREI III • Sidney Farber Cancer Institute, Boston, Massachusetts 02115.

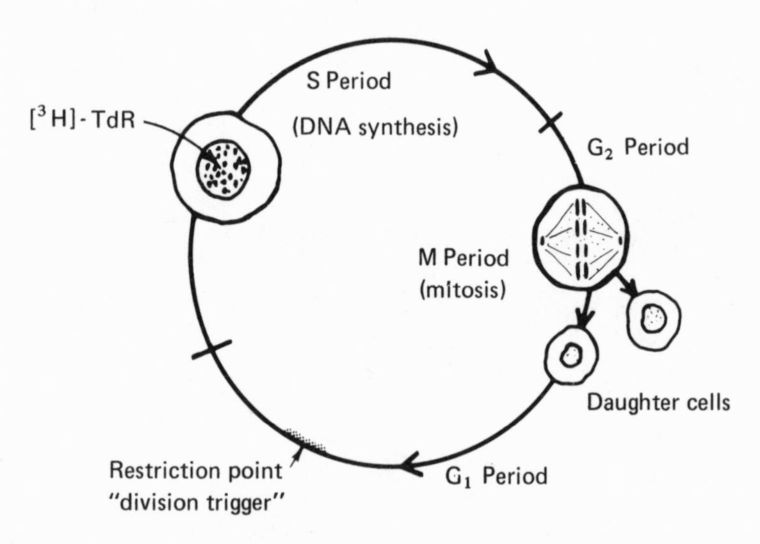

FIGURE 1. Mitotic cycle of the cell.

cancer cells follow this pattern. During the mitosis (M) period, the cell divides into two daughter cells. These proceed through the G_1 period. During the G_1 period, a crucial event may occur at the restriction or division trigger point that commits the cell to division. The cell then enters the DNA synthesis (S) period of the cycle. This phase of the cycle can be distinguished by autoradiography using the radioactive metabolite tritiated thymidine, which is an obligate precursor of DNA. There is a second brief gap known as the G_2 period between the end of DNA synthesis and the actual process of mitosis. Note that two phases of the cycle, the M period and the S period, can be distinguished by morphologic techniques.

If one exposes an asynchronous population of cells to tritiated thymidine, only those cells in the process of making DNA will be labeled (Figure 2). This cohort of cells will advance through mitosis and the analysis of the labeled mitosis

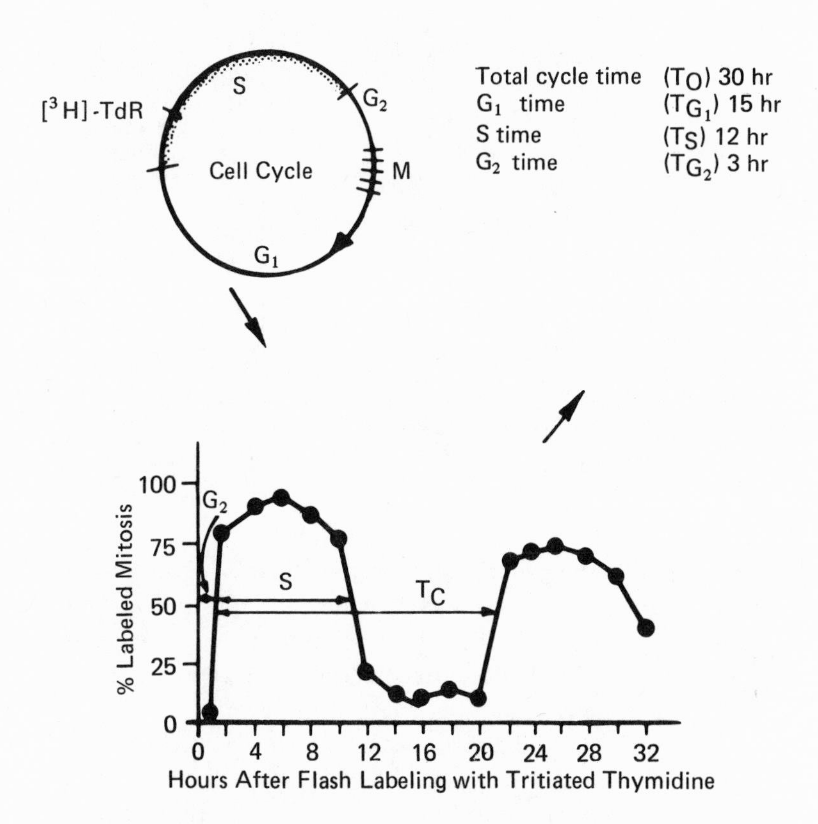

FIGURE 2. Labeled mitosis curve. S = DNA synthesis; [^{3}H]=TdR = tritiated thymidine.

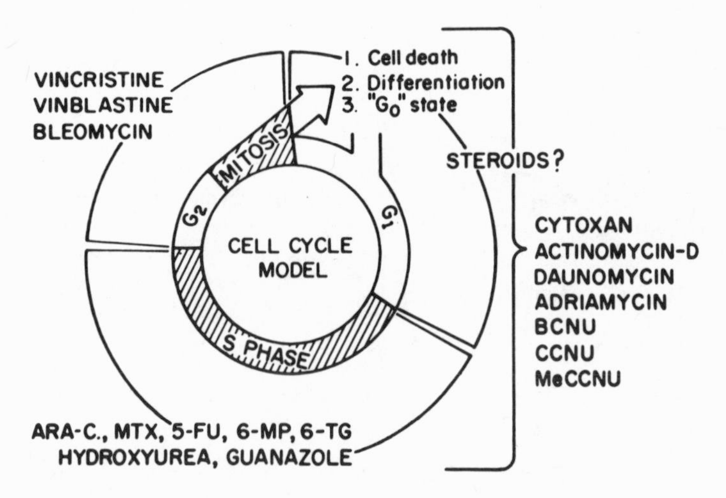

FIGURE 3. Cell cycle and its inhibition by antitumor agents. Compounds on the right inhibit cells during all stages of the cell cycle and may also inhibit noncycling cells.

curve, that is, the proportion of mitoses labeled with thymidine as a function of time may provide definitive information with respect to cytokinetics of the cell under study (Quastler and Sherman, 1959). Thus, the total cycle time, and the times of the G_1, S, and G_2 periods, may be determined.

Using specialized techniques, it has been possible to synchronize cells *in vitro* in various stages of the cell cycle and to study the effects of chemotherapeutic agents on these various stages (Figure 3). This has been extended to *in vivo* studies where, for example, vincristine, which arrests cells in mitosis, produces a marked increase in the proportion of cells in mitosis in its immediate wake. These cells leave mitosis with some degree of synchrony, which is evident by the increase in the number of cells making DNA a few hours later. This corresponds closely to the time required for cells to progress from the M period to the S period (Vadlamudi and Goldin, 1971; Frei *et al.*, 1964). Employing an agent such as ara-C, which destroys those cells in the process of making DNA, at this point in time could lead to a greater therapeutic effect. While a number of approaches involving such synchronization are underway, clinical progress has not yet been achieved, primarily because normal cells are similarly synchronized so that a therapeutic advantage is not achieved. Nevertheless, the use of cells synchronized in various stages of mitosis has provided important information on the cell cycle specificity of antitumor agents and has provided leads as to the mechanism of action of such agents.

When the above thymidine-labeled mitosis technique was applied to solid cancers in animals and man, several important cytokinetic concepts emerged (Figure 4). For example, in studies in patients with melanoma, it was found that the cell cycle time (T_c) was 3 days (Shirakawa *et al.*, 1970). Thus, if all of the cells were in cycle, the doubling time of the tumor would be 3 days. In fact, the volume doubling time was 42 days

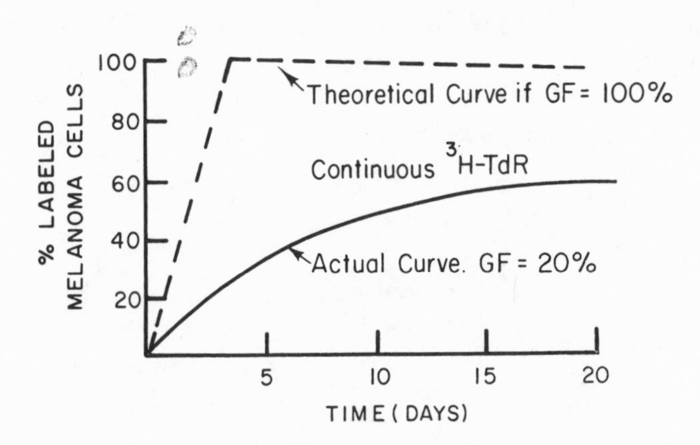

FIGURE 4. Growth fraction (GF) of melanoma in man. Mitotic cycle time = 3 days; volume doubling time = 42 days.

(Shirakawa *et al.*, 1970). Theoretically, if one administered tritiated thymidine by continuous infusion and the cell cycle time was 3 days, 100% of the tumor cells should label by 3 days. It was observed for solid tumors that substantially lower labeling index curves were almost invariable. This was explained by the fact that a substantial proportion of the cells within a tumor were not mitotically active, that is, were not growing. In fact, the growth fraction (GF, the proportion of mitotically active tumor cells over the total number of tumor cells) was often quite low, in the range of 5–20% (Mendelsohn, 1962). If such cells were end stage and therefore incapable of reentering the cycle, they would by definition not be malignant and would not be of concern to the therapist. However, in experimental solid cancers, it was found by syngeneic transplant and bioassay that over 80% of cells are capable of reentering the mitotic cycle from a tumor area wherein the growth fraction *in situ* was less than 5% (Figure 5).

Since most human cancers are associated with a low growth fraction, the importance of developing drugs capable of damaging tumor cells that are not in cycle has been emphasized. Most of the early drugs used for cancer, inhibiting DNA synthesis or the process of mitosis *per se,* were active primarily against cycling cells. Thus, they were effective against the more rapidly growing tumors such as the

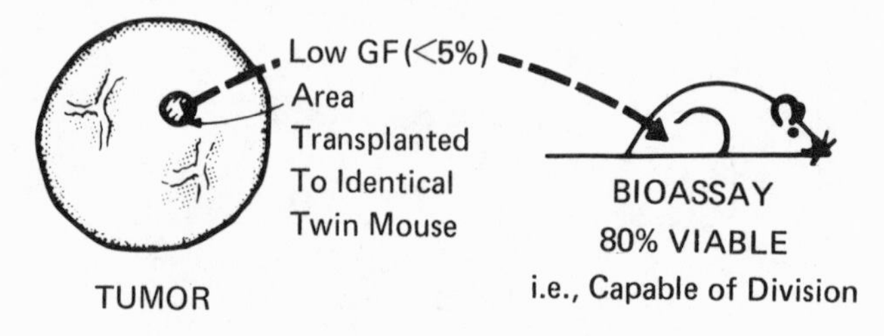

FIGURE 5. Viability of low growth fraction (GF) tumor cells. High GF areas are shaded; low GF areas are white.

leukemias and choriocarcinoma but not against the more slowly growing, low-growth-fraction, solid tumors. Accordingly, drug development was reoriented and test systems designed to discriminate drugs active against noncycling cells were developed. The comparative activity of several prototype and new antitumor agents against cells in culture, all of which are in cycle (the growth fraction is 100%), and against cells that are not in cycle, so-called resting cells, are presented in Table 1. It is apparent that the older agents are much less effective against noncycling cells, which require much higher concentrations to produce comparable inhibition. Emphasis on the identification and application of compounds which have activity against resting cells or low-growth-fraction tumors has led to the development of the nitrosoureas, DTIC, and adriamycin. These agents are almost as active against resting cells as compared to growing cells, and have been found to have major antitumor activity against such slow-growing tumors as breast cancer, brain tumors, melanoma, and lymphomas.

This has provided a major basis for the construction of combination chemotherapy. Thus, for low-growth-fraction tumors, agents active against cycling cells are combined with those active against noncycling or resting cells (Frei, 1972). An example of such is the so-called MOPP program for disseminated Hodgkin's disease (Table 2). This includes nitrogen mustard and procarbazine agents with substantial activity against noncycling cells, and prednisone. These agents are far superior when used in combination as compared to their independent use. They produce an increase in the proportion of patients who achieve complete disappearance of their tumor. More important, over 50% of patients so treated have remained tumor free for observation periods of up to 10 years. Thus, cytokinetic information has provided a major rationale for the selection and use of drugs in combination and has resulted in major progress in treatment, including curative

TABLE 1
Relative Effect of Antitumor Agents on "Growing" (Cycling) and "Resting" Tumor Cells

Drug	Drug concentration required to produce 50% inhibition of	
	Growing[a] cells	Resting[b] cells
Arabinosyl cytosine	1[c]	∞
Methotrexate	1	50
Fluorouracil	1	40
Vincristine	1	40
Adriamycin	1	7
DTIC	1	4
Nitrogen mustard	1	3
BC nitrosourea	1	0.5–2

[a]Log phase.
[b]Plateau phase.
[c]Arbitrarily set at 1 to emphasize ratio.

TABLE 2
Hodgkin's Disease: Combination Chemotherapy

Agent	Complete %	Remission duration (median in months)
M (nitrogen mustard)	10	5
O (oncovin, VCR)	13	4
P (prednisone)	<10	<4
P (procarbazine)	18	6
MOPP (combination chemotherapy)	80	>60

treatment of a variety of cancers in man (Frei, 1972; DeVita *et al.*, 1970).

Another cytokinetic concept which has helped treat cancer patients is the so-called first-order kinetics. A series of studies with the antitumor agent cyclophosphamide in the mouse leukemia L1210 is presented in Table 3. The treatment was constant and the variable was the number of tumor cells in the animal at the time treatment was initiated. With a large inoculin size, a very large number of tumor cells were killed and the fractional destruction of tumor cells was 99%. With differing numbers of tumor cells, the absolute number of cells killed varied enormously, but the fractional destruction of tumor cells remained constant. This law of first-order kinetics applies generally to biological and biochemical systems. It states that it is the fractional destruction of tumor cells by a given treatment that remains constant and independent of the number of tumor cells present. In view of this, it is important to appreciate that treatment which is capable of eradicating 99% of the cells would be mathematically, and is biologically, capable of curing small tumor burdens. While such treatment will produce tumor regression, it will not be curative for larger numbers of tumor cells. Such knowledge provided compelling evidence that the opportunity to cure a given tumor with chemotherapy is much greater if treatment is initiated when the minimum number of tumor cells is present (Skipper and Schabel, 1973).

Further evidence to this point is presented in Figure 6. All experimental solid tumors growing in rodents essentially follow this pattern. Thus, when the tumor is small the growth rate is exponential, but as it increases in size the growth rate decreases. This is evident in looking at the doubling time (DT) which for the small tumor is 4 days, but as the growth rate levels off exceeds 100 days. This decrease in growth rate is only slightly explicable on the basis of prolongation of the cycle time

TABLE 3
Percentage of Leukemic Cell Populations (10^2–10^6 Cells) Killed by a Given Dose of Cyclophosphamide[a]

Approx. number of leukemic cells inoculated i.v.	Average number of leukemic cells surviving treatment	Average number of leukemic cells killed	Average % leukemic cell population	Fraction of animals "cured" (zero surviving leukemic cells)
1,000,000	15,000	985,000	98.5	0/10
100,000	5,000	95,000	95.0	0/10
10,000	200	9,800	98.0	0/10
1,000	35	965	96.5	1/10
100	1	99	99.0	4/10

[a] The same single dose of cyclophosphamide (¼ of LD_{10} dose) was administered to all animals 24 hr after leukemic cell inoculation.

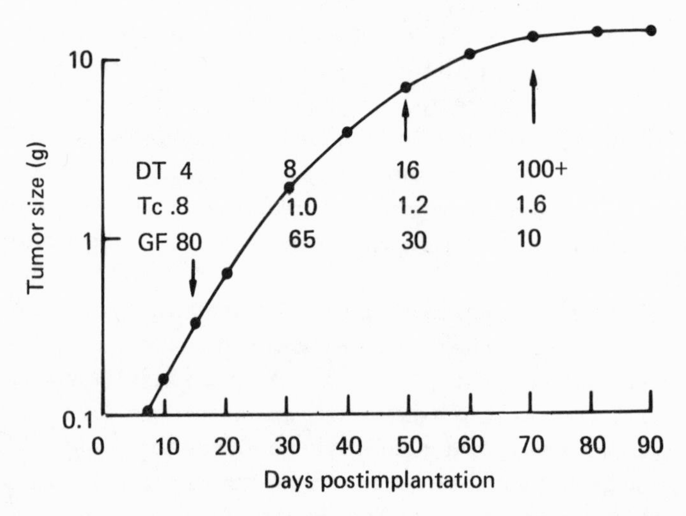

FIGURE 6. Experimental solid tumor kinetics. DT = doubling time (in days); GF = growth fraction (%); T_c = cell cycle time (in days).

(T_c) which changes from 0.8 to 1.6 days. The growth fraction (GF), on the other hand, decreases from 80% progressively down to less than 10%. Thus, as tumors increase in size, the proportion of cells in cycle markedly decrease (Simpson-Herren and Lloyd, 1970). Since our antitumor agents are most effective against cycling cells, it is advantageous to treat when tumors are small, not only because of the aforementioned first-order kinetics, but also because higher growth fraction tumors are more responsive to treatment. Thus, intensive treatment in patients in whom the bulk of the tumor mass has been controlled by complete remission induction chemotherapy, or by surgery and/or X-ray, becomes important.

This latter approach has been referred to as adjuvant chemotherapy and one circumstance where it has proven effective is breast cancer (Figure 7). Stage II breast cancer, by definition, includes spread of the tumor to lymph nodes in the armpit. Local control of the tumor can be achieved by surgery and/or radiotherapy. The major obstacle to cure in premenopausal Stage II breast cancer is the fact that microscopic deposits of tumor have spread beyond the local area and will become clinically apparent in such sites as the lung, liver, and bones, such that by 3 years, 30% of patients with one to four nodes and 80% of patients with more than four positive nodes have relapsed (see the control curves in Figure 7). If, however, such patients are treated with combination chemotherapy (cyclophosphamide, methotrexate, fluorouracil) for 12 months, the relapse rate is significantly reduced (Bonadonna *et al.*, 1976). Other adjuvant chemotherapy programs, similar in principle but differing in detail, have been successful in other tumors such as bone tumors in children (Jaffe *et al.*, 1974).

The above cytokinetic factors that led to the major progress achieved with combination chemotherapy and with adjuvant chemotherapy derived from cell cycle studies that were first performed several years ago. More recently, our knowl-

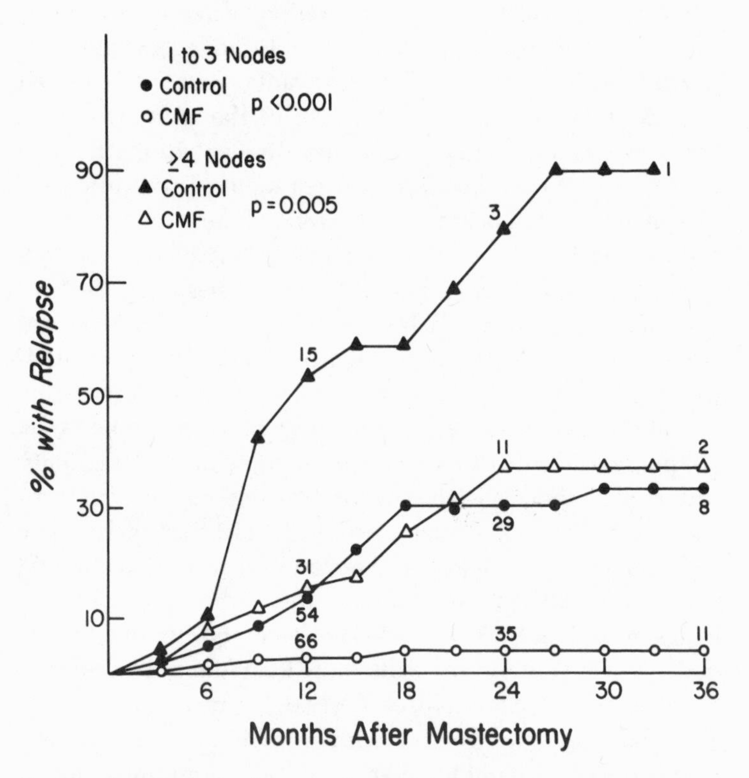

FIGURE 7. Effect of adjuvant chemotherapy on Stage II premenopausal breast cancer.

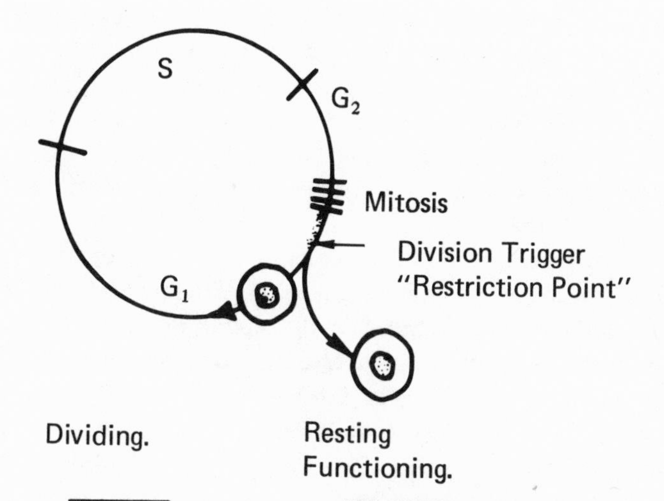

Uncontrolled expansion of cell population (cancerous cell)	Dividing only on signal to provide growth or cell renewal (normal cell)

Division trigger always on "GO" for cancer cell.

1. What is division trigger?
2. What are the controlling (intrinsic and extrinsic) factors?
3. How do they differ for cancer cell?
4. How can this knowledge be used for the benefit of the patient?

FIGURE 8. Cancer cell division: the division trigger is always on "go" for the cancer cell.

edge concerning the cell cycle has expanded very substantially and perturbations of the cell cycle by various cancer treatment modalities are increasingly understood at a biological and biochemical level. Such information is being employed, and the construction of current clinical trials and further improvement in the treatment of patients with cancer may be anticipated.

Finally, it is important to consider the fundamental question which the tumor biologist must ask with respect to the cell cycle. In the G_1 period, there is a division trigger point (Figure 8) or restriction point that determines whether the cell will proceed through the cycle and divide, or whether it will drop out of the cycle and rest. Cancer cells, by definition, are propelled into division in an uncontrolled fashion. Normal cells, on the other hand, are resting and functioning; having dropped out of the cycle, they will divide only upon normal demands for growth or cell renewal. Why is the division trigger always on "go" for the cancer cell? What is the division trigger? How is it controlled and how does it differ in the cancer cell as compared to the normal cell? Hypotheses and research addressed to this central issue by Pardee and others are becoming increasingly sophisticated. In addition to its basic science implications, such knowledge would most certainly lead to applied advances in biomedicine, generally and for the cancer patient particularly.

REFERENCES

Bonadonna, G., Brusamolino, E., Valagussa, P., Rossi, A., Brugnatelli, L., Brambilla, C., DeLena, M., Tancini, G., Bajetta, E., Musemuci, R., and Veronesi, U., 1976, Combination chemotherapy as an adjuvant treatment in operable breast cancer, *New Engl. J. Med.* **294:**405.

DeVita, V., Serpick, A., and Carbone, P., 1970, Combination chemotherapy in the treatment of advanced Hodgkin's disease. *Ann. Intern. Med.* **73:**881.

Frei III, E., 1972, Combination cancer chemotherapy, *Cancer Res.* **32:**2593.

Frei III, E., Whang, J., Scoggins, R. B., Van Scott, E. J., Rall, D. P., and Ben, M., 1964, The stathmokinetic effect of vincristine, *Cancer Res.* **24**:1918.

Jaffe, N., Frei III, E., Traggis, D., and Bishop, Y., 1974, Adjuvant methotrexate and citrovorum-factor treatment of osteogenic sarcoma, *New Engl. J. Med.* **291**:994.

Mendelsohn, M. L., 1962, Autoradiographic analysis of cell proliferation in spontaneous breast cancer of C3H mouse. III. Growth fraction, *J. Natl. Cancer Inst.* **28**:1015.

Quastler, H., and Sherman, F. G., 1959, Cell population kinetics in the intestinal epithelium of the mouse, *Exp. Cell Res.* **17**:420.

Shirakawa, S., Luce, J. K., Tannock, I., and Frei III, E., 1970. Cell proliferation in human melanoma, *J. Clin. Invest.* **49**:1188.

Simpson–Herren, H. E., Lloyd, H., 1970, Kinetic parameters and growth curves for experimental tumor systems, *Cancer Chemother. Rep.* **54**:143.

Skipper, H. E., and Schabel, F. M., Jr., 1973, Quantitative and cytokinetic studies in experimental tumor models, in: *Cancer Medicine* (J. Holland and E. Frei III, eds.), pp. 629–650, Lea & Febiger, Philadelphia.

Vadlamudi, S., and Goldin, A., 1971, Influence of mitotic cycle inhibitors in the antileukemic activity of cytosine arabinoside (NSC-63878) in mice bearing leukemia L1210, *Cancer Chemother. Rep.* **55**:547.

How Research on the Cell Biology of Reproduction Can Contribute to the Solution of the Population Problem

DON W. FAWCETT

Mankind is now confronted by two inseparable problems of awesome magnitude—the uncontrolled growth of the world population and a serious shortage of food for the undernourished millions already born and for the millions yet to be born. The attainment of zero population growth in some regions of this country has led to a regrettable complacency about the population problem. But the fact remains that some 200,000 persons are being added to the world population every day—74 million new mouths to feed every year. The United Nations has identified 43 countries with very low per capita income, inadequate diet, and large food deficits. It is especially distressing that the highest fertility rates are found in many of those same developing countries that are struggling to im-

DON W. FAWCETT • Hersey Professor of Anatomy, Harvard University Medical School, Cambridge, Massachusetts 02138.

prove the lot of their people only to find that unchecked population growth is wiping out the gains achieved in agricultural production. Research on the cell biology of reproduction provides the understanding of basic mechanisms that is essential for progress in the limitation of human fertility and in the promotion of fertility in food-producing domestic animals.

The cell that is the "villain" in the drama of overpopulation is the spermatozoon, but this same cell can be a "hero" in the struggle to increase beef and dairy production. The study of its development, its structure, and its function has already yielded rich economic dividends for animal husbandry and unexpected benefits for human medicine.

Some 25 years ago, British cell biologists Alan Parkes and Audrey Smith were studying the effects of low temperature on cells. They worked with bull spermatozoa because these were readily available as free cells in suspension and their motility provided a useful indicator of cell viability. They discovered that the common chemical glycerol, in appropriate concentrations, protected spermatozoa against the lethal effects of freezing. Although these investigators had no practical objective in mind in undertaking these experiments, their work rapidly led to the commercial preservation of bull sperm by freezing at −196°C for subsequent use in artificial insemination. The first calf born as a result of artificial insemination was affectionately called Frosty I (Figure 1). He was to be followed by millions of others.

This advance made possible the international shipment of frozen semen to improve the quality of cattle in developing nations. Economy was achieved in that the number of dairy bulls maintained in any country could be greatly reduced. Few of our children in the Northeast have ever seen a bull. More poignant is the fact that very few dairy cows in New England have ever seen a bull either, but they calve regularly. A million and a half cows in the Northeast are bred annually with frozen

FIGURE 1. Photograph of Frosty I, the first calf born of artificial insemination with bull semen that had been preserved by freezing at −70°C. (From C. Polge and L. E. Rowson, *Nature* **169,** 1952.)

semen from bulls at the Eastern Artificial Insemination Unit in Ithaca, New York. The economic benefits in the past 25 years of the discovery of the cryoprotective effects of glycerol would be difficult to estimate. But of even greater importance to mankind is the fact that the knowledge gained from these pioneering experiments, later made possible the preservation of blood cells for transfusion, and the long-term storage of valuable tissue culture cell lines used both in cancer research and in the production of vaccines for protection against several human infectious diseases.

In the 1950s, it was discovered independently by Austin and by Chang that spermatozoa upon discharge still have not attained their full fertilizing capacity. They undergo further physiological changes in the female reproductive tract that make them competent to penetrate the envelopes around the ovum in order to achieve fertilization. This final preparation for their reproductive mission is called *capacitation*. There is now abundant evidence that this is a prerequisite for conception in most mammals, possibly including the human. The precise nature of the change involved in sperm capacitation is still a subject of active investigation by cell biologists, but methods have been found for reproducing the phenomenon *in vitro*. Recognition of the need for capacitation, the discovery of methods for inducing maturation of ova, and the development of chemically defined media for the maintenance of ova in tissue culture stimulated efforts to achieve fertilization outside the female reproductive tract. By 1960, full-term rabbits had been obtained from ova fertilized *in vitro* and transferred to the uterus of recipient mothers. Subsequent research has made *in vitro* fertilization of ova from the common laboratory species a routine procedure and has provided an experimental system that will make it possible now to investigate the molecular mechanisms involved in mammalian fertilization.

Here again the insight gained from fundamental studies at the cellular level is being applied to problems of great potential importance to food production. Although *in vitro* fertilization has not yet been accomplished for all of the large domestic animals, there is no doubt it soon will be. In the meantime, the technology of egg transfer to a host animal is already well advanced. It has been possible for more than a decade to recover a fertilized ovum from a donor cow and successfully transfer it to a recipient foster mother that serves as an incubator, carrying the embryo through gestation to live birth of a normal calf. Two fertilized ova have been introduced nonsurgically into the uterus of a recipient cow resulting in the birth of twins. The host animal has no effect on the coat color or appearance of the offspring, which depends of course on the genetic characteristics of the donor. Thus, normal calves of two different breeds have been carried to term simultaneously in the two uterine horns of a foster mother of a third breed. More recently, a normal bull calf was born in Cambridge, England, after the transfer to a foster mother of a blastocyst that had been deep frozen for some time at −196°C (Figure 2). The calf, Frosty II, is now 2 years old, growing normally and in good health. Several lambs have also been born after transfer of deep-frozen embryos. It is too soon to foresee all the commercial benefits to mankind of these recent developments but they will probably be great.

On the other side of the ledger, there have been many advances that may ultimately contribute to the development of new methods for limiting human fertility. We have learned much in recent years about the events in fertilization and have identified vulnerable steps in the process that may lend themselves to pharmacological interruption.

When the ovum is released from the ovary at ovulation, it is enclosed in a thick glassy membrane, the *zona pellucida,* and this in turn is invested by several layers of adhering ovarian follicular cells collectively called the *cumulus oophorus* (Figure 3A). An intriguing issue has been how the spermatozoon traverses these barriers to reach the egg for fertilization.

FIGURE 2. Photograph of Frosty II, the first calf born as a result of implantation in a host cow of a blastocyst that had been deep frozen at −196°C. The foster "mother" is black: the genetically unrelated calf is a white-faced auburn-colored Hereford. (From Wilmut and Rowson, *Vet. Rec.* **92:**686, 1973.)

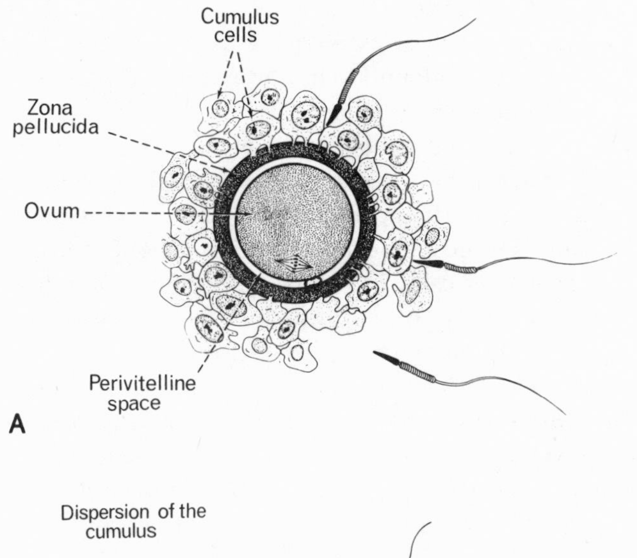

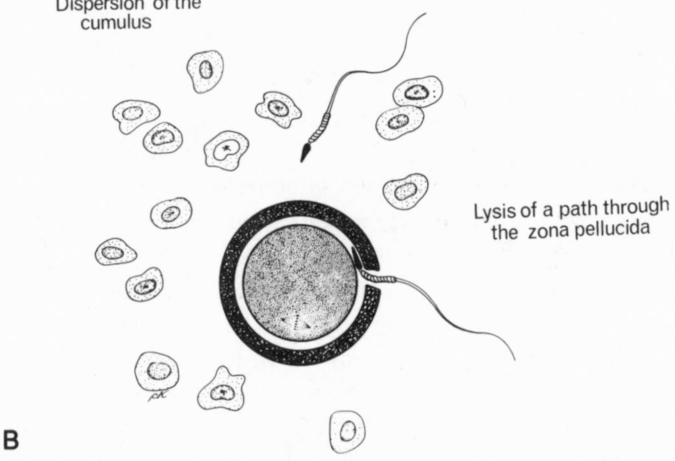

FIGURE 3. A: Drawing of a recently ovulated mammalian ovum surrounded by its zona pellucida and the adherent cumulus cells. The spermatozoa must traverse these barriers and reach the perivitelline space to carry out fertilization. B: The enzymes released by the acrosome of the spermatozoon result in dispersion of the cumulus cells and lysis of a path through the zona pellucida that permits penetration of the spermatozoon into the perivitelline space.

Studies with the electron microscope have provided clues to the mechanism. In electron micrographs, the anterior part of the sperm nucleus is found to be covered by a cap-like structure called the *acrosome* (Figure 4). This has been investigated with great interest by cell biologists because it appears to be essential for sperm penetration. It was formerly thought to be a structure unique to the spermatozoon but it now seems likely that it is a specialized form of an organelle that is of widespread occurrence among cells. In the past 15 years, nearly all cell types in the body have been found to contain small membrane-enclosed organs called *lysosomes.* These contain a variety of hydrolytic enzymes that enable the cell to digest foreign material taken into it, and in some circumstances these digestive enzymes may be released from cells into the extracellular space. Lysosomes can be identified in living cells by their capacity to take up and concentrate the fluorescent dye, acridine orange.

When living spermatozoa are exposed to acridine orange, the acrosomal cap fluoresces brilliantly. Thus, it shares one of the identifying characteristics of lysosomes. Moreover, if live spermatozoa are spread upon a microscope slide that has been coated with gelatin, a clear halo soon appears around each sperm head due to digestion of the adjacent gelatin by a proteolytic enzyme released from the acrosome. A similar release of enzymes is thought to be a cardinal feature of the mechanism of sperm penetration. Although several enzymes have been extracted from sperm acrosomes, two that are believed to be of greatest physiological significance are *acrosin,* a proteolytic enzyme, and *hyaluronidase,* an enzyme that digests the carbohydrate-rich extracellular substance hyaluronic acid. Electron microscopic studies of fertilizing sperm have shown us how the release of enzymes takes place.

When capacitated spermatozoa encounter a recently ovulated egg in the upper reaches of the oviduct and begin to

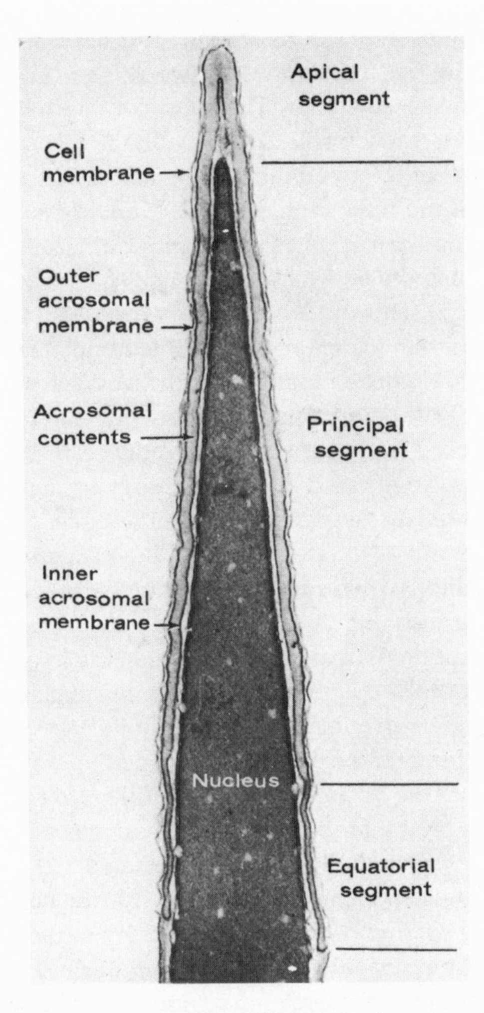

FIGURE 4. Electron micrograph of longitudinal section through the head of a rhesus monkey spermatozoon, showing the condensed chromatin of the nucleus and the overlying acrosomal cap. The acrosome is limited by inner and outer acrosomal membranes. The latter is closely invested by the plasmalemma of the spermatozoon. (From Bloom and Fawcett, *Textbook of Histology,* W. B. Saunders Company, 1975.)

insinuate themselves among the cumulus cells, some emanation from the egg evidently initiates the so-called *acrosome reaction* of the spermatozoa. The nature of the chemical signal triggering the reaction still eludes us. In the responding spermatozoa, the outer membrane of the acrosomal cap contacts and fuses at multiple sites with the overlying surface membrane of the sperm head, creating numerous openings through which the enzyme-rich contents of the acrosome escape (Figure 5). This process of membrane fusion rapidly progresses until the outer acrosomal membrane and the cell membrane are reduced to myriad small vesicles which dissociate, leaving the anterior part of the sperm head covered only by the exposed inner acrosomal membrane. The hyaluronidase and acrosin released in this process bring about the loosening and ultimate dispersion of the cumulus cells (Figure 3B). Residual acrosin associated with the inner acrosomal membrane is credited with lysing a path through the zona pellucida for the vigorously motile spermatozoon.

The traditional concept of fertilization that envisioned the pointed end of the sperm head entering the egg like an arrow has proved to be erroneous. Once within the narrow perivitelline space, the broad side of the flattened sperm head rests gently against the tips of minute finger-like projections of the egg surface (Figure 6). Where these microvilli come into contact with the region of the sperm head immediately behind the acrosome, their membrane coalesces with the surface membrane of the spermatozoon. The continuity of the sperm and egg membrane is rapidly extended from these sites of initial coalescence and the sperm head, and ultimately the tail as well, sink into the egg cytoplasm. The sperm nucleus decondenses and fuses with the egg nucleus to complete fertilization by union of the genetic complement of the two parents. Cell biologists are now devoting much investigative attention to the special properties of the postacrosomal region of the sperm

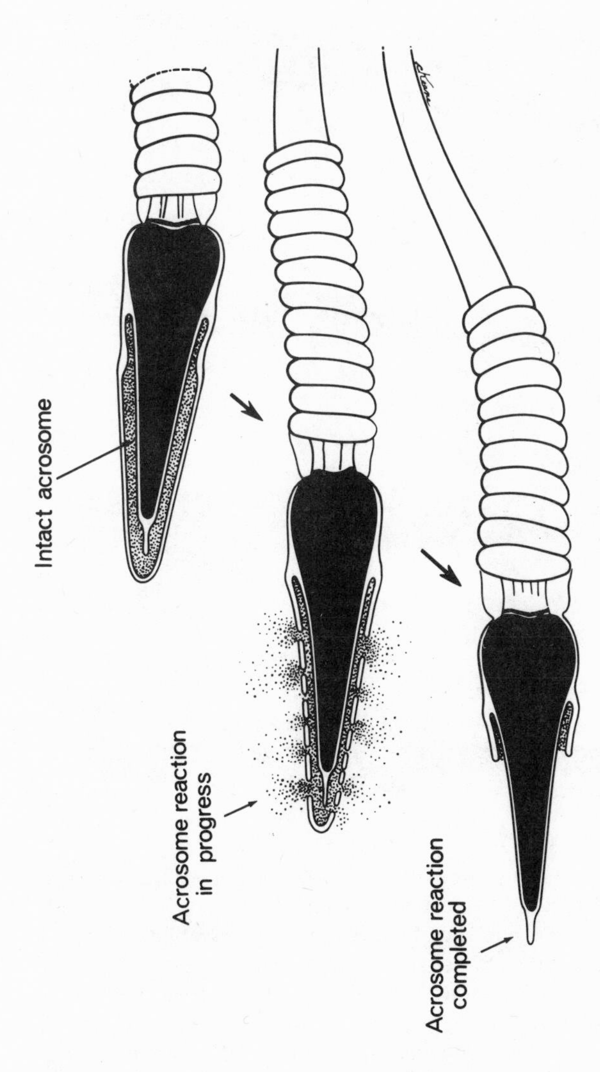

FIGURE 5. Diagram of the acrosome reaction. Triggered by some unknown factor, the outer membrane of the acrosome fuses at many points with the overlying cell membrane creating openings through which the enzyme-rich contents of the acrosome escape. At the completion of the acrosome reaction, the anterior part of the head is covered only by the inner acrosomal membrane. In this condition, it penetrates the zona pellucida.

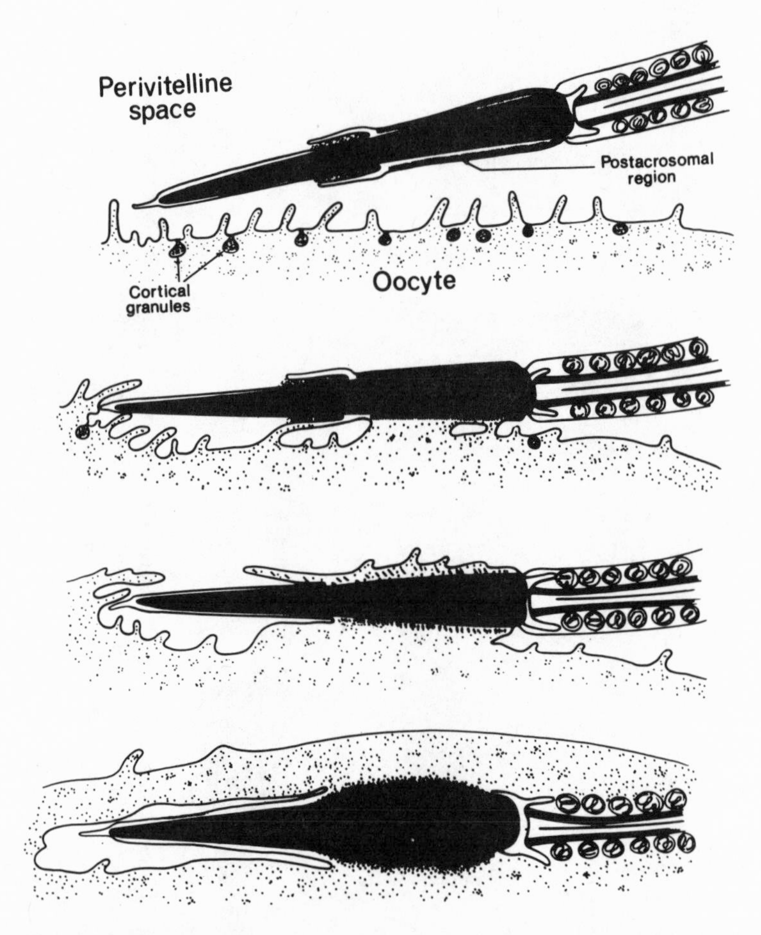

FIGURE 6. Schematic representation of the events in fertilization. The postacrosomal region of the spermatozoon contacts and fuses with the surface membrane of the egg. This triggers secretion of cortical granule material which is believed to alter the zona pellucida, preventing other sperm from entering the perivitelline space. The sperm head slowly sinks into the egg cytoplasm and its nucleus undergoes decondensation to form the male pronucleus. (Redrawn after R. Yanigamachi, in *Regulation of Mammalian Reproduction,* Charles C Thomas, 1973.)

cell membrane that are responsible for recognition of the egg and fusion with its membrane—properties which no other region of the sperm surface possesses. An understanding of the basis for recognition might suggest means of blocking it for purposes of contraception.

Of the hundreds of millions of sperm introduced into the female at copulation, only one enters the egg. The mechanism by which others are excluded has long been one of the intriguing problems of reproductive biology. Although still unsolved, we are able to focus the search more sharply. There are many small dense bodies called *cortical granules* in the peripheral cytoplasm of the ovum when continuity is established between the membrane of the egg and that of the foremost spermatozoon. This event triggers secretion of the content of the cortical granules into the perivitelline space. The observation that the zona pellucida of a fertilized egg is more resistant than that of an unfertilized egg to the action of proteolytic enzymes *in vitro* has suggested that substances released from the cortical granules may alter the physical properties of the zona so as to make it more resistant to penetration by a second sperm. There is much more research to be done here but it seems clear that if we could simulate the natural mechanism of the block to polyspermy, we might have a physiological means of denying entry to the first.

The problem of sperm locomotion is of central importance to population control for there is no fertility without sperm motility. Since Anton Leeuwenhoek first described the eel-like swimming movements of spermatozoa 300 years ago, biologists have been fascinated by the mechanism of flagellar motion. Early microscopists could not imagine how the incredibly slender and seemingly structureless sperm tail could generate and propagate waves of bending along its length. Much of the mystery remains, but in recent years cell biologists, using the electron microscope to achieve magnifications up to

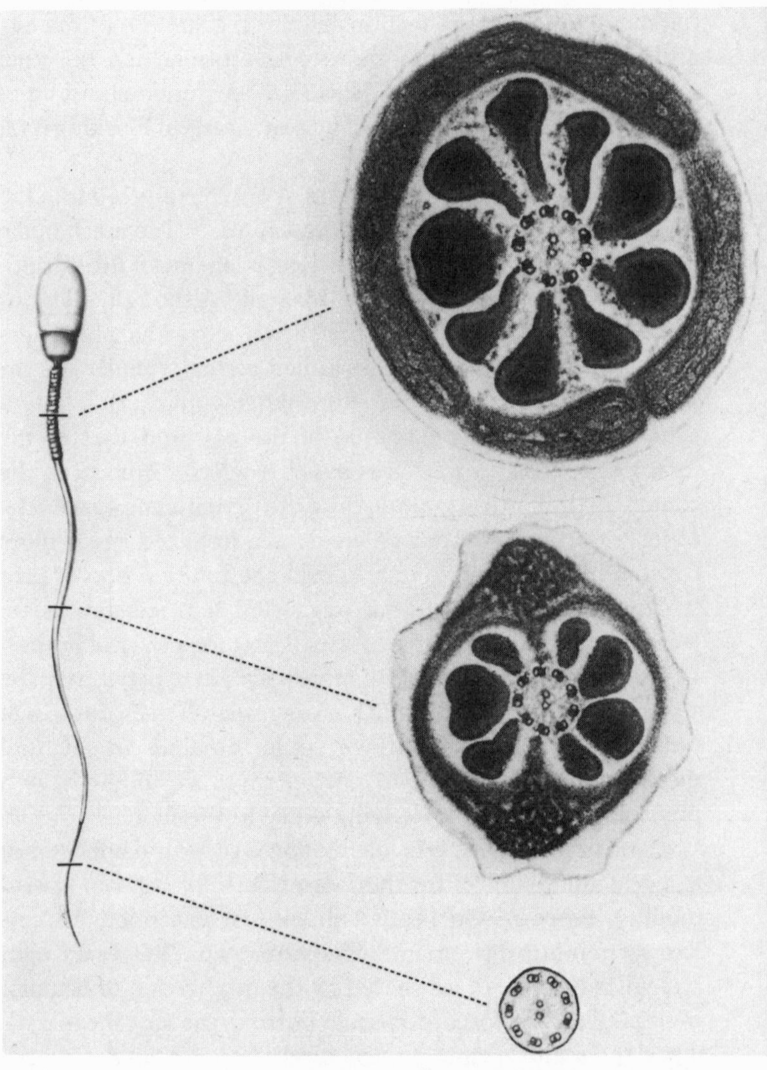

FIGURE 7. The appearance of electron micrographs of cross sections of the sperm tail at the level of the middle piece with its mitochondrial sheath (*top*), the principal piece with its fibrous sheath (*middle*), and the end piece in which there is no sheath other than the cell membrane (*bottom*).

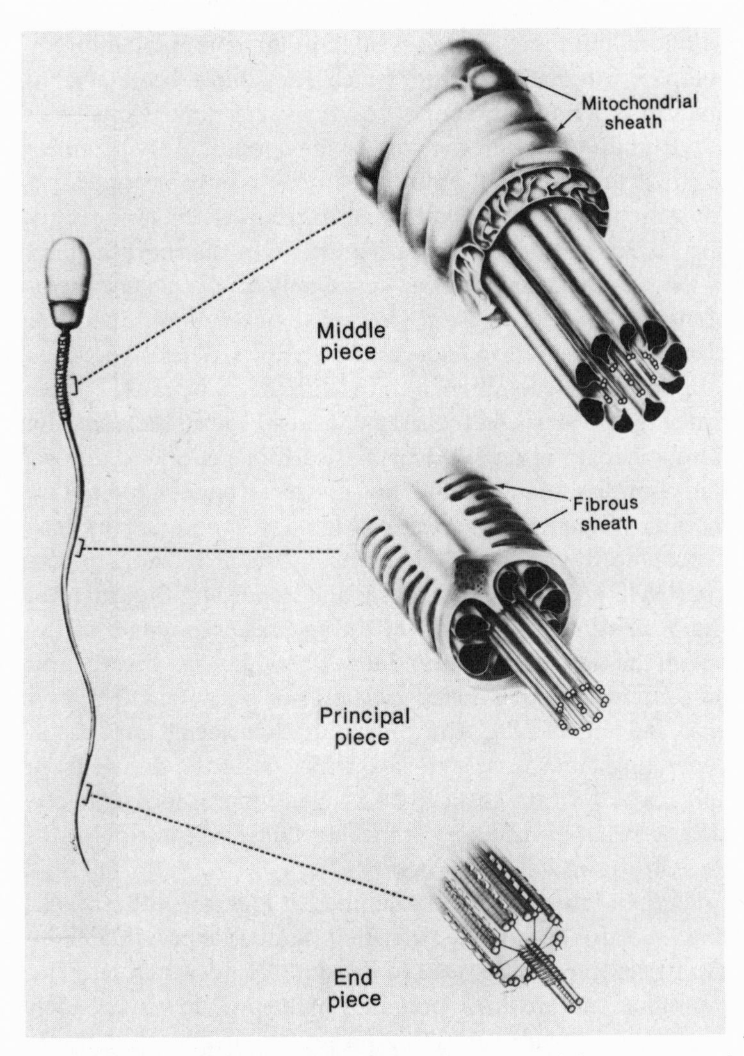

FIGURE 8. Extended three-dimensional diagrams of the structure of the three regions of the sperm tail. In the middle piece, the axoneme and outer dense fibers are enclosed by an energy-generating mitochondrial sheath. In the principal piece, this is replaced by a fibrous sheath, and in the end piece the axoneme is enclosed only by the cell membrane.

half a million times, have revealed in the sperm tail a highly complex internal structure which they have been able to analyze down to components of molecular dimensions.

Running through the core of the sperm tail for its entire length is the axoneme, a bundle of microtubules arranged in a constant pattern with a central pair surrounded by nine evenly spaced doublet microtubules (Figure 7). In the short *end piece* of the tail, the axoneme is enclosed only by the plasma membrane; in the long *principal piece,* nine outer dense fibers are placed outside the doublets and a complex fibrous sheath is interposed between them and the surface membrane. Nearer the base of the tail, in the segment called the *middle piece,* the fibrous sheath is replaced by a sheath of helically disposed mitochondria that provide the chemical energy for sperm motility (Figure 8). Modern cell biology progresses by first describing the structural elements and their relations in the intact cell, and then by isolating and describing the chemical characteristics of each. All of the major components of the sperm tail—axoneme, outer dense fibers, fibrous sheath, and mitochondria—have been isolated in pure fractions and analyzed chemically. The only part that seems capable of generating movement is the axoneme. The outer dense fibers and fibrous sheath seem to be resilient scleroproteins maintaining structural integrity of the flagellum and contributing to its stiffness and elastic properties.

When the axoneme is examined at high magnification in cross sections (Figure 9), a constant number of protofilaments can be resolved in the wall of the doublet microtubules. The protofilaments are linear polymers of the protein *tubulin.* Each doublet has a pair of *arms* that project toward the next one in the row. The arms are spaced at regular intervals along the length of the microtubules. The detailed structural analysis of the axoneme was not sufficient to establish how it might generate movement, but careful observation of the disposition of microtubules at the tips of cilia in different phases of the beat

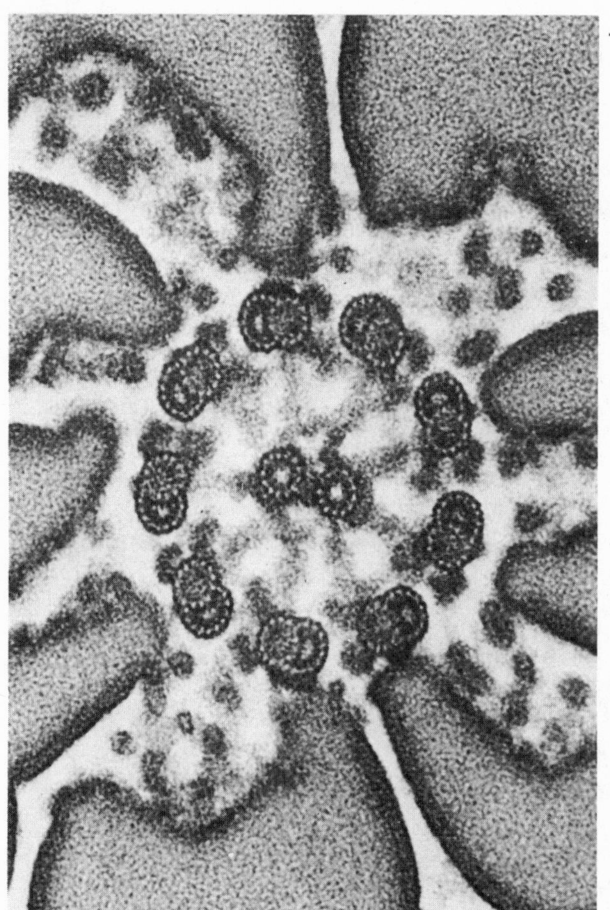

FIGURE 9. High-magnification electron micrograph of the axoneme of the sperm tail stained with tannic acid and lead citrate. Protofilaments of tubulin can be seen in negative image in the wall of the single and doublet microtubules. Each doublet has a pair of arms projecting clockwise. (Micrograph by Dr. Dan Friend.)

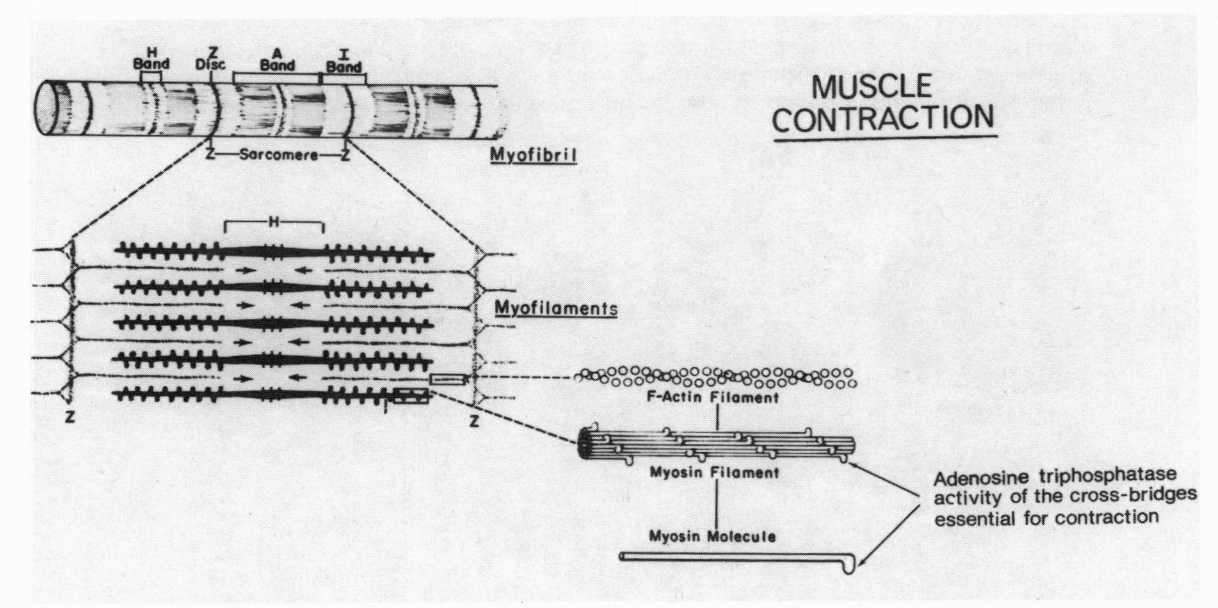

FIGURE 10. Diagram of the interdigitating arrangement of thick and thin filaments of skeletal muscle. Shortening results from sliding of the thin actin filaments toward the middle of the sarcomere. The active agents of the sliding are short cross bridges on the thick filaments which represent the heads of the myosin molecules. These have adenosine triphosphatase activity which is necessary to provide energy for contraction.

enabled Peter Satir (1968, 1974) to establish that the doublet microtubules do not shorten during bending. It was suggested therefore that by analogy with skeletal muscle, the microtubules might slide with respect to one another. The myofibrils of skeletal muscle are composed of interdigitating sets of thin (actin) filaments and thick (myosin) filaments which interact with one another via minute cross bridges (Figure 10). When muscle shortens, activation of the bridges results in longitudinal displacement of the actin filaments which penetrate from opposite ends deeper into the interstices between the myosin filaments. The projecting cross bridges are the flexible heads of the myosin molecules that make up the thick filaments. They contain the enzyme adenosine triphosphatase (ATPase), which is responsible for conversion of chemical energy into the mechanical work of muscle shortening. The occurrence of regularly spaced projecting arms on the doublets of the flagellar axoneme (Figure 11), made it tempting to specu-

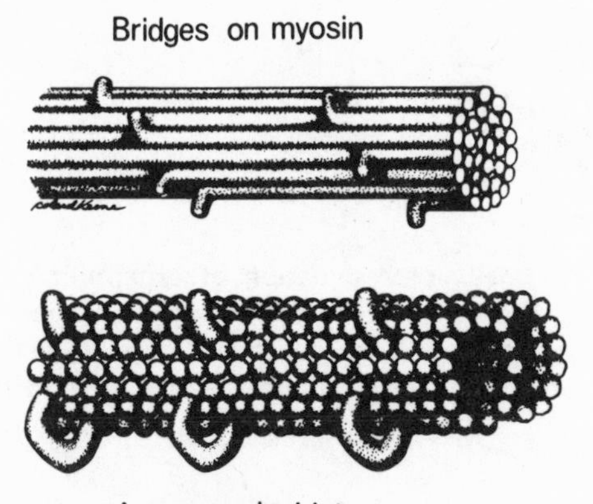

FIGURE 11. Regularly recurring "arms" on subfiber A of the doublet microtubules in cilia and flagella and "bridges" on the myosin filaments of muscle. The arms of the doublets suggest the possibility that they are functionally analogous to the bridges of the myosin filaments.

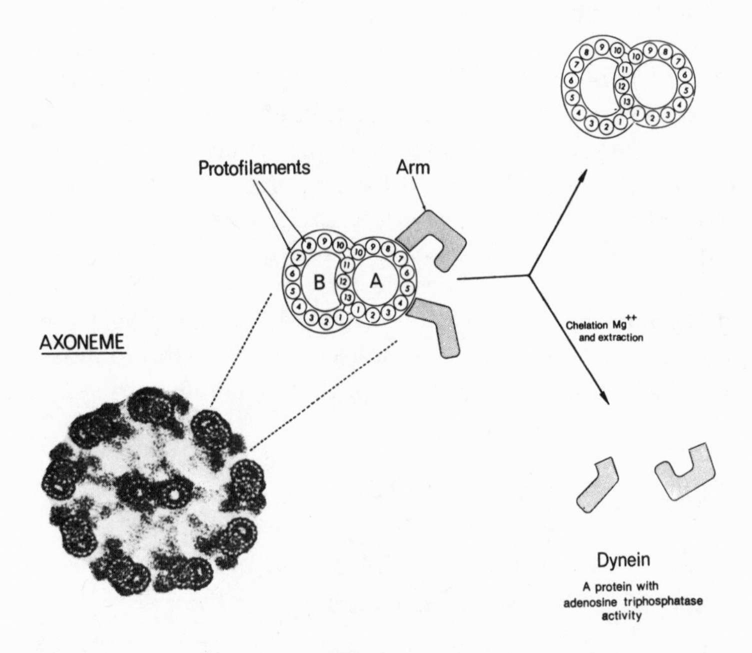

FIGURE 12. Schematic representation of experiments by Gibbons (1965) in which the arms were extracted from the axonemal doublets and found to consist of a large protein molecule with adenosine triphosphatase activity. This protein was named dynein.

late that these might be analogous to the bridges on myosin filaments and might play a comparable role. It remained to be demonstrated, however, that the arms contain ATPase and that the doublet microtubules of the axoneme do slide.

It was therefore a major advance in the chemical dissection of the sperm flagellum and in our understanding of its motility when Ian Gibbons (1965) showed that the arms on the doublets can be selectively extracted (Figure 12). They consist of a large protein molecule to which he gave the name *dynein* and it does, in fact, have adenosine triphosphatase activity. With Summers (Summers and Gibbons, 1971), he went on to demonstrate microtubule sliding in a highly ingenious experiment. The membrane was removed from segments of sperm flagella by exposing them to a detergent, and then they were subjected to mild proteolytic digestion to free the doublets from their attachment to other components of the axoneme (Figure 13). When adenosine triphosphate was added to this preparation as an energy source, the segments of axoneme were observed in the microscope to elongate rapidly as shearing forces created by interaction of the doublets caused them to slide and move out opposite ends of the tail fragments (Figure 14). Strong evidence was thus obtained for a sliding-tubule hypothesis of flagellar motility implicating the dynein arms on the doublets in the transduction of chemical energy to the mechanical work of sperm locomotion.

This fundamental research in cell biology which nonscientists might consider trivial or criticize as the idle pursuit of intellectual curiosity at public expense has proved to have unexpected clinical relevance. Henning Pedersen in Denmark (Pedersen and Rebbe, 1975) and Björn Afzelius (1976) in Sweden have now studied several men who sought medical advice for an unusual form of infertility in which normal numbers of spermatozoa are produced but none are motile. When examined with the electron microscope, these spermatozoa

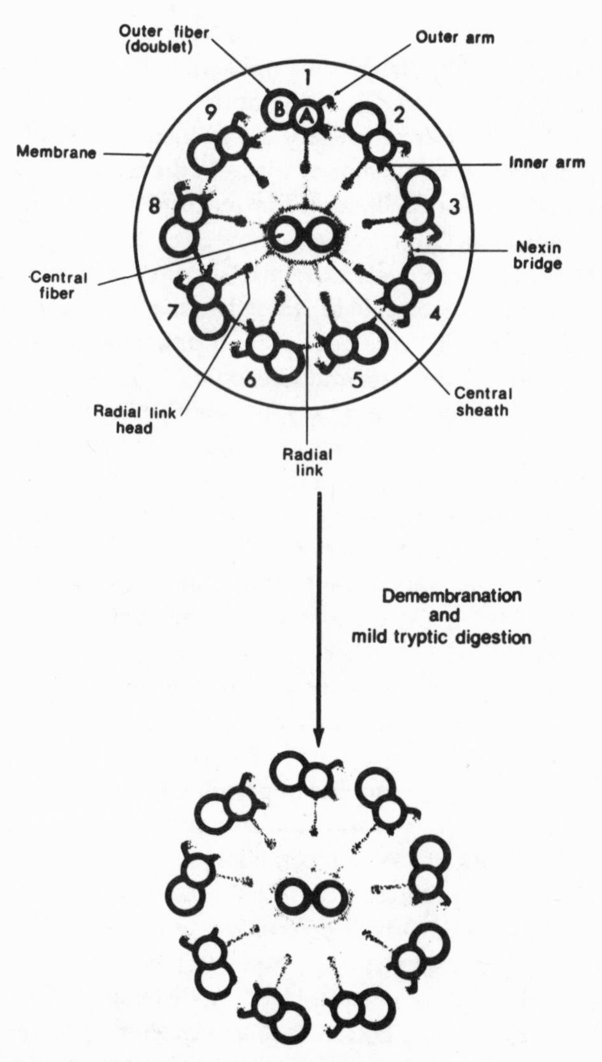

FIGURE 13. Doublets of the axoneme, which are attached to each other by nexin links and to the central pair by radial spokes. As a preliminary to testing the capacity of doublets to slide on addition of ATP, segments of flagella were first demembranated in detergent and briefly digested with trypsin to free the doublets from their attachments.

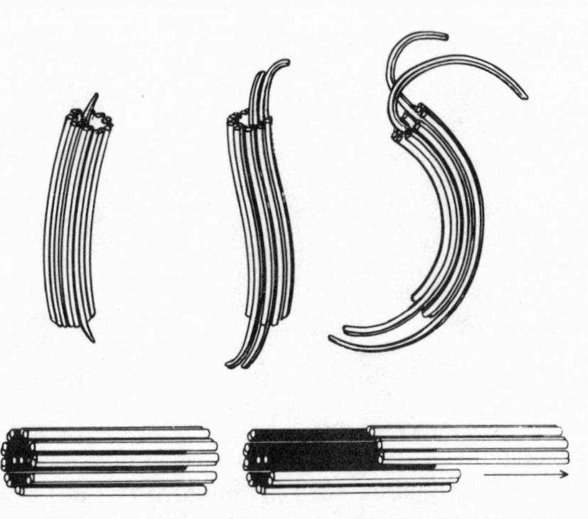

FIGURE 14. Schematic representation of an experiment supporting a sliding tubule mechanism of flagellar motility. Demembranated segments of sperm tails treated with ATP rapidly elongated as a result of sliding of doublets out opposite ends. (Redrawn from Satir, *Scientific American* **231**:44, 1974.)

have essentially normal ultrastructure except for the absence of arms on the doublet microtubules of the axoneme (Figure 15). These individuals are evidently genetically incapable of synthesizing dynein. Further study of these patients revealed that in addition to infertility, they also suffer from chronic sinusitis and bronchiectasis because, lacking dynein, the cilia in their respiratory tract are also immobile and they are unable to move out of the bronchi and paranasal sinuses the blanket of mucus that normally traps and conveys inhaled bacteria and inert particulate matter into the nasopharynx where they can be eliminated by swallowing. These patients share another abnormality that remains unexplained. They have situs inversus viscerum, a developmental anomaly of symmetry in which the heart and other organs are located on the wrong side. The association of the clinical findings of chronic sinusitis, bronchiectasis, and situs inversus viscerum was long recognized by clinicians under the name of Kartagener's syndrome, a disease

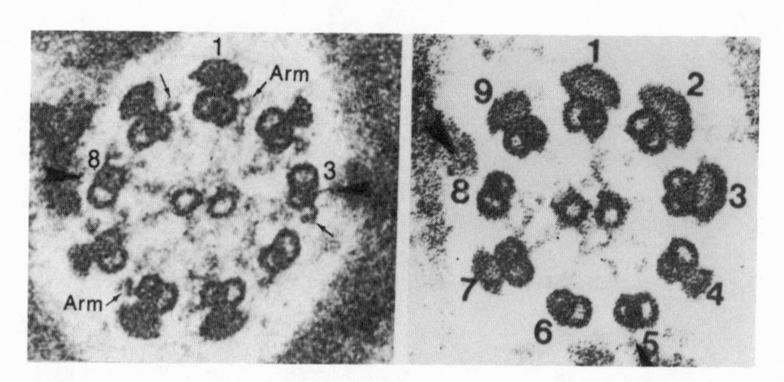

FIGURE 15. Electron micrograph of the axoneme of a normal human spermatozoon (left) in which arms are clearly visible on each doublet of the axoneme, compared with spermatozoon from a man with Kartagener's syndrome (right) where the dynein arms are missing. (From Pedersen and Rebbe, *Biol. Reprod.* **12:**541, 1975.)

of unknown etiology. Basic research on cell biology has provided an explanation for the molecular and genetic basis of two of the triad of clinical signs and has added one previously unrecognized, namely male infertility. At the same time the study of these patients has provided a confirmation of the essential role of dynein in ciliary and flagellar motility. Continued progress in understanding the mechanism of sperm motility may ultimately contribute to the rational development of safe antifertility drugs for the male. In the meantime, we work toward that goal with the empirical approach of testing for antifertility effects drugs that are synthesized for other purposes.

Spermatozoa are produced in astronomical numbers in a sixth of a mile of convoluted seminiferous tubules in each testis (Figure 16). When they leave the testis, they are not yet capable of progressive motility and are unable to fertilize ova. They are moved on from the testis into a highly convoluted duct some 16 ft long that forms the *epididymis,* an organ adherent to the posterior surface of the testis. During their slow passage

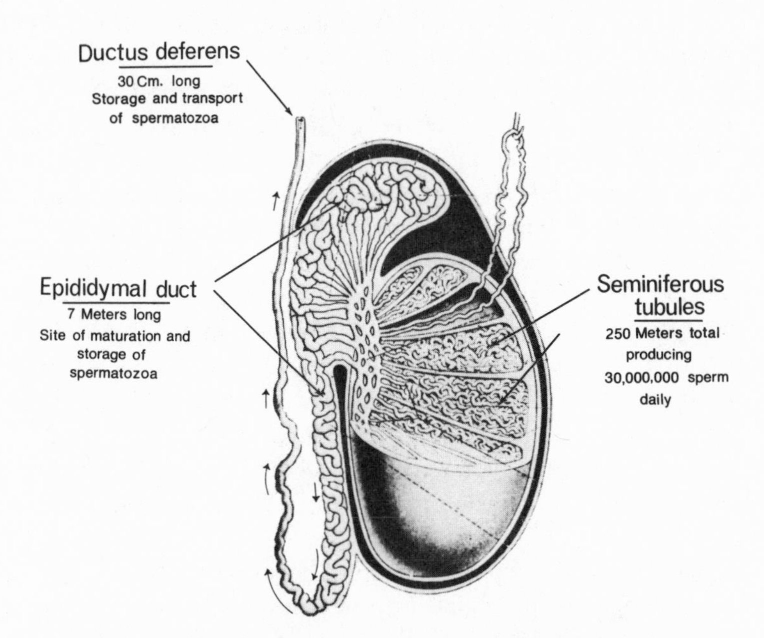

FIGURE 16. The structure of the human testis and its excurrent ducts. Current efforts to find a safe, reversible, oral contraceptive for the male are concentrated on the epididymal duct where it is hoped maturation of stored sperm might be arrested so that they would not acquire the capacity to fertilize. (Modified by permission after Hamilton *et al.*, *Textbook of Human Anatomy*, 1956.)

through the epididymal duct, the spermatozoa mature, gradually acquiring the potential for normal motility and fertility. The epididymis is now the subject of intensive research designed to define its biochemical functions and to find drugs that are concentrated in its lumen and which might act on the spermatozoa to prevent their maturation. This stratagem seems to hold considerable promise. Two drugs have been found that produce reversible infertility in experimental animals by an action at the level of the epididymis: alpha chlorohydrin and its analogue 1-amino-3-chloro-2-propanol appear to act upon an essential enzymatic step in the energy-generating mechanism of the spermatozoa. A third, dichlorbenzyl indazol carboxylic acid, induces an immobilizing deformity of the sperm tail. Unfortunately, the toxicity of these compounds for primates precludes human trials but there is reason to be hopeful that other compounds with a posttesticular antifertility effect will be discovered which are free of undesirable side effects. The prospects for early success would be greatly enhanced by more basic research to establish the biochemistry and physiology of the cells lining the epididymal duct.

It has only been possible in this brief essay to present a few examples of the potential contribution of cell biology to the amelioration of the population problem. I have directed my remarks to the spermatozoon and the male tract because the need for intensified effort is greatest there. Research on the reproductive biology of the male has lagged many years behind that on the female. We are now in a period of significant progress in our understanding of the hormonal control of spermatogenesis, the mechanism of sperm release and of sperm maturation, the process of fertilization, and the physiology of the excurrent duct system. A number of potentially vulnerable steps in these processess have been identified. The state of our knowledge of the male may soon be comparable to that which preceded the development of oral contraceptives for the female. If adequate research support can be sustained, it is reasonable to expect that the next 20 years will see the development of safe, effective means of fertility control in the male.

Suppression of fertility is not the only objective of such research, nor will it be the only reward. Greater understanding of male reproductive physiology will improve our capacity to help those couples who earnestly desire a child but are unable to have one. And as pointed out at the beginning of this chapter, much of the fundamental research that has been done on the male was motivated by a desire to improve the fertility of domestic animals. The benefits in the production of food and fiber for a growing population have been great and their impact will continue as improved methods of animal breeding spread to developing nations. These indirect returns must also be taken into account in any assessment of the societal benefits of research on the cell biology of reproduction.

REFERENCES

Afzelius, B. A., 1976, A human syndrome caused by immotile cilia (Kartagener's syndrome), *Science* **193**:317.

Bloom, W., and Fawcett, D. W., 1975, *Textbook of Histology*, W. B. Saunders, Philadelphia.

Fawcett, D. W., 1975a, The mammalian spermatozoon, *Dev. Biol.* **44**:394.

Fawcett, D. W., 1975b, Gametogenesis in the male: Prospects for its control, in: *Developmental Biology of Reproduction* (C. L. Markert and J. Papaconstantinou, eds.), pp. 25–53, Academic Press, New York.

Gibbons, I. R., 1965, Dynein, a protein with ATPase activity from cilia, *Science* **149**:424.

Hamilton, W. J. (ed.), 1956, *Textbook of Human Anatomy*, MacMillan, London.

Mann, T., 1975, Animal reproduction and artificial insemination, *Proc. R. Inst. G.B.* **48**:107.

Pedersen, H., and Rebbe, H., 1975, Absence of arms in the axoneme of immobile human spermatozoa, *Biol. Reprod.* **12**:541.

Polge, C., and Rowson, L. E. A., 1952, Fertilizing capacity of bull spermatozoa after freezing at −79°C, *Nature* (London) **169:**626.
Satir, P., 1968, Studies on cilia. III. Further studies on the cilium tip and a "sliding-filament" model of ciliary motility, *J. Cell Biol.* **39:**77.
Satir, P., 1974, How cilia move, *Sci. Am.* **231:**44–63.
Smith, A. U., 1961, *Biological Effects of Freezing and Supercooling,* Edward Arnold, London.
Summers, K. E., and Gibbons, I. R., 1971, Adenosine triphosphate induced sliding of tubules in trypsin treated flagella of sea-urchin sperm, *Proc. Natl. Acad. Sci. USA* **68:**3092.
Wilmut, I., and Rowson, L. E. A., 1973, Experiments on the low temperature preservation of cow embryos, *Vet. Rec.* **92:**686–690.
Yanigamachi, R., 1973, Behavior and functions of the structural elements of the mammalian sperm head in fertilization, in: *Regulation of Mammalian Reproduction: Proceedings* (S. J. Segal, R. Crozier, P. A. Corfman, and P. G. Condliffe, eds.), National Institutes of Health Conference, Charles C Thomas, Springfield, Ill.

How Basic Studies of Insects Have Helped Man

CARROLL M. WILLIAMS

"THE BUGS ARE COMING" exclaims the banner headline of a recent cover story in *Time* magazine. In the article itself we learn how the human race is presently losing ground in its age-old battle with the insects. The scenario involves two intractable difficulties. The first is that conventional chemical pesticides which are being used throughout the world on an ever-increasing scale are too broad in their effects. They are toxic not only to the pests at which they are aimed but also to other animals. Moreover, by persisting in the environment, and sometimes even increasing in concentration as they are passed along the food chain, they present a hazard to other organisms including man. The second difficulty is that insects have shown a remarkable ability to evolve a resistance to conventional chemical pesticides. For this reason the conventional pesticides—our first line of defense against insect pests and vectors of disease—are losing much of their former effectiveness.

CARROLL M. WILLIAMS • Benjamin Bussey Professor of Biology, Harvard University, Cambridge, Massachusetts 02138.

There is nevertheless reason to believe that the realities of the situation may not be quite so gloomy. Since World War II, governmental agencies such as the NSF, the NIH, and the USDA have fostered basic research on the biology, physiology, and endocrinology of insects. The same is true for certain private organizations such as the Rockefeller Foundation and more recently for certain branches of the chemical industry. A by-product of these investigations has been the identification of weak spots in the physiological and ecological armor of insects—Achilles' heels that can be attacked by what have been termed "biorational" techniques.

I propose to illustrate how basic research has enhanced our potential for controlling insects by reviewing the story of the hormonally active "third-generation pesticides." Let us first note that insects, just like vertebrate organisms including human beings, make use of specific blood-borne chemical messengers to control and coordinate their growth, metabolism, behavior, and reproduction. These signaling agents are the hormones. But it is important to understand that the detailed chemistries of the insect hormones are very different from those of vertebrates. They are the products of more than 300 million years of separate evolution. Consequently, the insect hormones, although vastly active on insects, are inactive when administered to mammals and other vertebrates. The same inactivity is seen when mammalian hormones are administered to insects.

As in the case of higher organisms, the insects make use of a complicated system of interacting hormones. And like higher organisms, the highest center in the insect endocrine system proves to be located in the brain itself. Here the cell bodies of certain nerve cells are able to synthesize and secrete a number of different hormones. The hormones exit from the brain by traveling along the axons of the neurosecretory cells to the so-called corpora cardiaca and corpora allata where they can be released into the blood. One of the brain's neurosecretions has the ability to activate a pair of endocrine organs called the prothoracic glands located on each side of the insect's anterior end. Under this stimulation the prothoracic glands synthesize and secrete a hormone called ecdysone. The latter circulates in the blood and is taken up by the cells throughout the body to cause developmental reactions.

When ecdysone acts unopposed, it provokes metamorphosis, a change illustrated by the transformation of a caterpillar into a pupa and the latter into a butterfly. Metamorphosis is typical of virtually all insects. It may be noted, however, that the so-called lower insects (grasshoppers, roaches, the true bugs, etc.) omit the pupal stage. The mature larva responds to ecdysone by transforming directly to the adult condition (see Figure 1).

But the lifestyle of insects is not one of ceaseless change and revolution. Under normal conditions the larva (caterpillar) postpones metamorphosis until it attains a certain critical size. To accommodate this growth the external (cuticular) part of the insect skin is molted from time to time after a new and larger cuticle has been secreted.

It is easy to show that the prothoracic gland hormone ecdysone is the prime mover in provoking the successive larval molts. All this suggests that there must be some sort of conservative force which modifies the cellular reaction to ecdysone, thereby preventing the metamorphosis which would otherwise take place.

This conservative force proves to be yet another endocrine agent, the so-called juvenile hormone. The hormone is synthesized and secreted by a pair of previously mentioned head glands, the corpora allata (Figure 2). The latter are continuously active in the secretion of juvenile hormone until the conclusion of the larval period. If one surgically removes the corpora allata so that an immature larva is deprived of juvenile

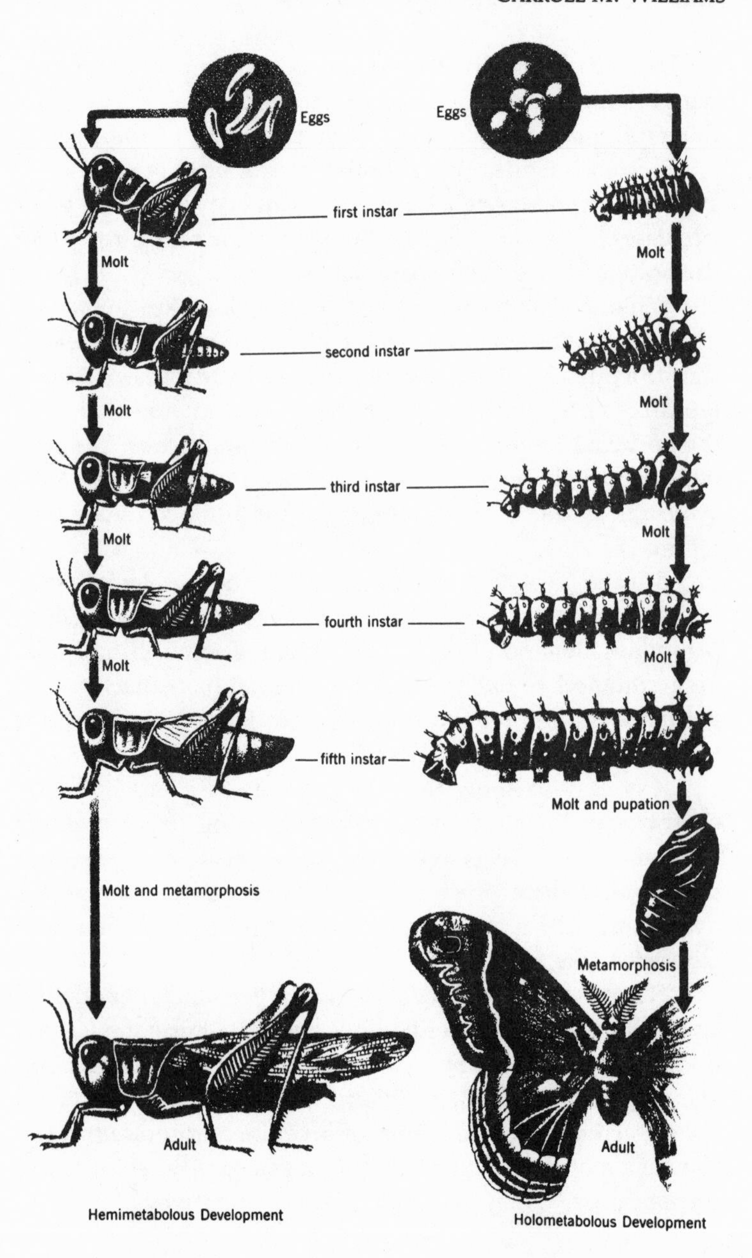
Eggs
Eggs
first instar
Molt
Molt
second instar
Molt
Molt
third instar
Molt
Molt
fourth instar
Molt
Molt
fifth instar
Molt and pupation
Molt and metamorphosis
Metamorphosis
Adult
Adult
Hemimetabolous Development
Holometabolous Development

hormone, it undergoes precocious metamorphosis to form a miniature pupa or adult which soon dies without further development.

From this résumé we see that larval growth requires that ecdysone and juvenile hormone act side by side. In order for metamorphosis to take place, the corpora allata must be turned off and the secretion of juvenile hormone must stop. The decision as to whether the corpora allata will secrete or not secrete juvenile hormone is apparently made by the brain. When the larva attains a certain critical size, the brain turns off the corpora allata as a necessary prelude to metamorphosis. This influence seems to be transmitted to the corpora allata via the nerves that connect them with the brain.

For many years after its discovery the juvenile hormone remained something of a will-o-the-wisp, resisting all efforts to extract or obtain it apart from the living insect. Then, quite by accident, a rich depot of the hormone was found to be present in the abdomens of male cecropia moths. This was indeed a fortuitous discovery for from that day to this only the closely related male cynthia and gloveri moths have been found to accumulate juvenile hormone in this way. We originally thought that the hormone was stored in all the abdominal tissues of male cecropia moths. Subsequently, it was found to be sequestered in the so-called accessory sex glands of the male moth. Why these glands take up and store so much hormone in the case of the male cecropia moth remains a mystery.

←

FIGURE 1. Diagrams showing hemimetabolous development in the grasshopper and holometabolous development in the cecropia silkworm. Holometabolous development is characterized by the intervention of an intermediate pupal stage between larva and adult; it is characteristic of the so-called higher insects including all moths, flies, beetles, and bees. (From A. Gorbman and H. A. Bern, *Textbook of Comparative Endocrinology*, John Wiley and Sons, 1966.)

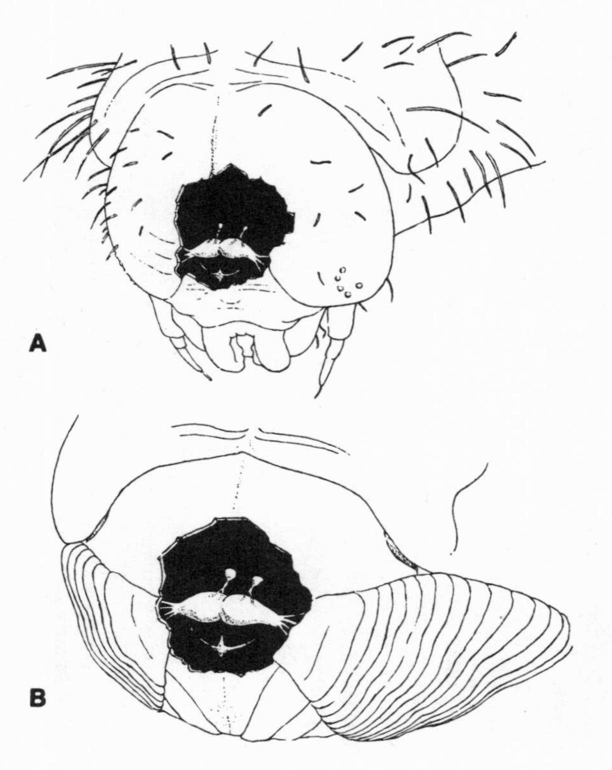

FIGURE 2. The brain and corpora allata of the cecropia silkworm are shown in cutaway views of (A) the head of the larva; (B) the pupa, and (C) the adult moth. The corpora allata are the two small bodies connected to the back of the brain by four tiny nerves containing the axons of the brain's neurosecretory cells. (From C. M. Williams, *Sci. Am.*, February 1958.)

Be that as it may, by extracting the abdomens with ether, it was possible to obtain a golden oil which showed high juvenile hormone activity when bioassayed. The crude extract was able to duplicate all of the effects previously realized when juvenile hormone was supplied by the implantation of living, active corpora allata. For example, when injected into a pupa just prior to the initiation of adult development, it caused the formation of bizarre creatures that showed a mixture of pupal and adult characters. All such animals soon died without completing adult development.

Indeed, it was soon found that the extract need not be injected. It sufficed to place the golden oil in contact with the unbroken skin through which it promptly penetrated to exert its effects. The same was true when the cecropia extract was tested on many different species of insects.

The study up to that point had been sufficiently mission unoriented to scandalize any congressional committee. All this changed overnight with the realization that here was a material which on contact with living insects was able to derail the normal course of metamorphosis. In the initial report of the successful extraction of the hormone it was possible to suggest that:

> In addition to the theoretical interest of the juvenile hormone, it seems likely that the hormone, when identified and synthesized, will prove to be an effective insecticide. This prospect is worthy of attention because insects can scarcely evolve a resistance to their own hormone.*

The next step was to isolate the hormone and establish its chemical identity. Several laboratories attacked this problem utilizing as a starting point the golden oil obtained from thousands of male cecropia moths. To make a long story short, three variants of the hormone were finally isolated and characterized. As shown in Figure 3, all three proved to be the methyl esters of the epoxide of certain unsaturated fatty acid derivatives. Since they show only minor differences from one an-

*From C. M. Williams, 1956, The juvenile hormone of insects, *Nature* **178**:213.

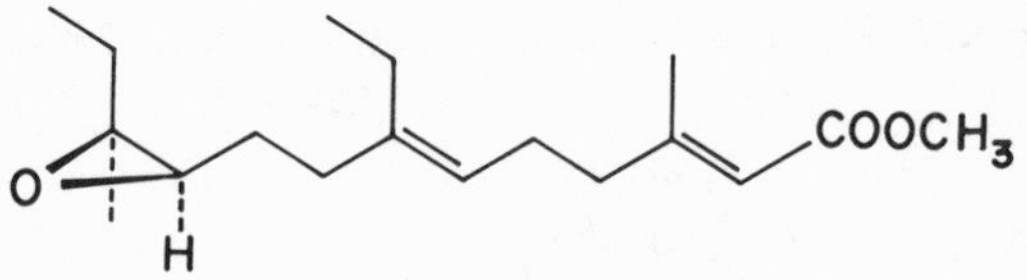

$COOCH_3$

C_{18} JH or JH I

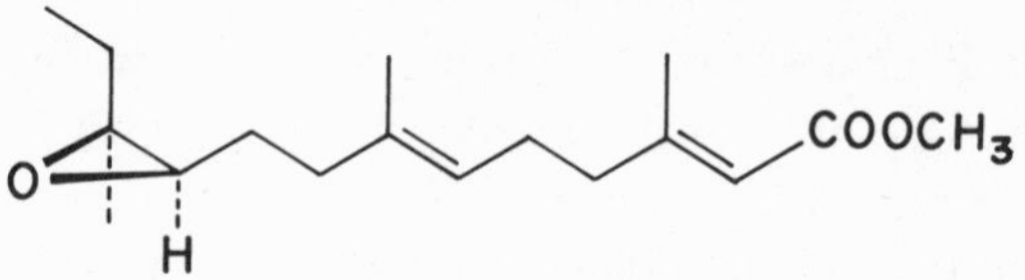

$COOCH_3$

C_{17} JH or JH II

$COOCH_3$

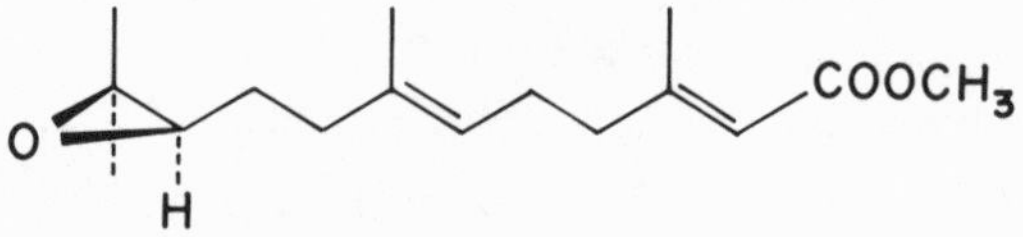

C_{16} JH or JH III

FIGURE 3. The three authentic juvenile hormones, which show only minor differences from one another.

other, they are called JH I (containing 18 carbon atoms), JH II (containing 17), and JH III (containing 16). All three were synthesized and tested for their ability to block the metamorphosis of diverse species of insects.

Under laboratory conditions, JH I and JH II were found to be extremely active when tested on most species of insects; JH III was nearly always several orders of magnitude less active. But when the highly active compounds were tested under conditions simulating field conditions, the results were very disappointing due to their extreme biodegradability. This fact, coupled with the difficulty and high cost of their synthesis, ruled out the use of any of the authentic juvenile hormones as pesticides.

Using the authentic hormones as models, the chemical industry undertook the synthesis of hundreds of cheaper and more stable analogs whose activities were bioassayed on various species of insects. A surprising finding was that the juvenile hormone mimics often show high activity for certain kinds of insects and low activity for others. That being so, the mimics can, in principle, be tailor-made to attack individual pests.

A case in point is the analog Altosid (Methoprene, or ZR–515; see Figure 4), which under laboratory conditions showed great effectiveness in blocking the metamorphosis of mosquitoes (see Figure 5) and certain other diptera, but little activity for most other insects. A special microencapsulated, slow-release formulation of Altosid (SR–10) has recently re-

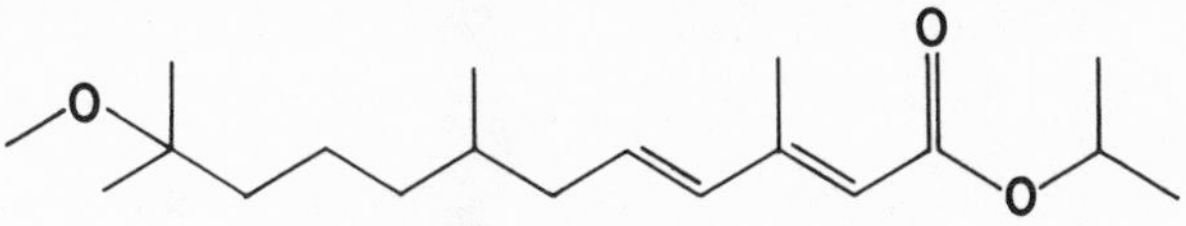

Methoprene (Altosid®; ZR–515)

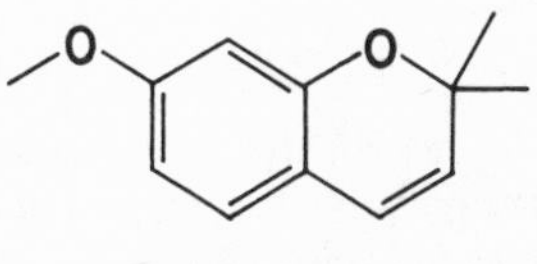

Precocene I

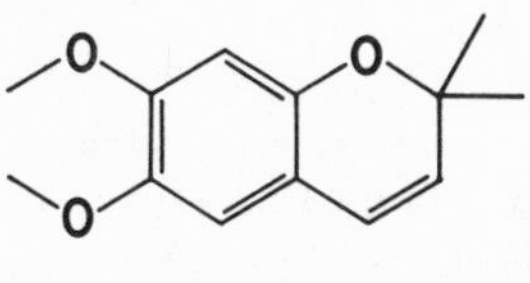

Precocene II

FIGURE 4. Methoprene, the first juvenile hormone analog to receive full registration for the control of mosquitoes; and the two precocenes, the first antijuvenile hormones discovered by W. S. Bowers *et al.* (W. S. Bowers, T. Ohta, J. S. Cleeve, and P. A. Marsella, 1976, Discovery of insect anti-juvenile hormones in plants, *Science* **193:**542–547).

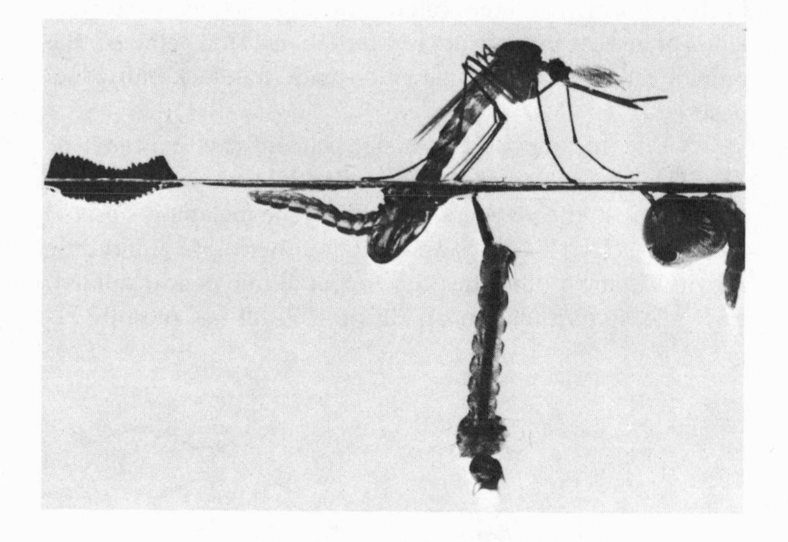

FIGURE 5. A waterline view of the metamorphosis of a mosquito. Each egg in the floating egg mass (*extreme left*) gives rise to a larva (*center, breathing through tube*). The larva transforms into a pupa (*extreme right*) and the pupa into the adult mosquito. The latter is shown emerging from the old pupal skin. (From A. J. Nicholson, 1931, *Bull. Ent. Res.* **22**:307.)

ceived a full United States commercial registration for use in controlling the floodwater mosquito *Aedes nigromaculus*—a species that had become resistant to virtually all conventional pesticides. Commercial marketing began in 1975. Over 100,000 acres, principally in California and Florida, were successfully treated. Good control is said to have been obtained by the application of as little as 0.3 ounce per acre. A further EPA registration has been sought for a new formulation of Altosid that promises to be sufficiently stable to use in the selective control of other genera and species of mosquitoes.

An ingenious strategy is the addition of yet another slow-release formulation of Altosid (CP–10) to salt licks and granular mineral supplements used in animal husbandry. In this "feed-through" procedure, enough Altosid remains unmetabolized when passing through the cattle's digestive tract to block the metamorphosis of manure-feeding houseflies and hornflies. This solution of a hitherto intractable problem promises to save millions in terms of livestock values throughout the world. In May 1975, the EPA issued full registration for this purpose.

Altosid is the first of what may hopefully prove to be a whole battery of third-generation pesticides aimed at specific insect pests and vectors of disease. It is worth emphasizing that those third-generation pesticides which are juvenile hormone analogs act to derail metamorphosis. In this manner they prevent any damage the adults might do while simultaneously blocking the pest's ability to reproduce the next generation of offspring. Nevertheless, the juvenile hormone analogs leave unanswered the control of contemporary generations of immature larval insects. This is a very considerable disability in the case of many agricultural pests where the larvae are often the most damaging stage in the life history.

Here again, basic research has provided what may prove to be a solution to this problem. We have already seen that

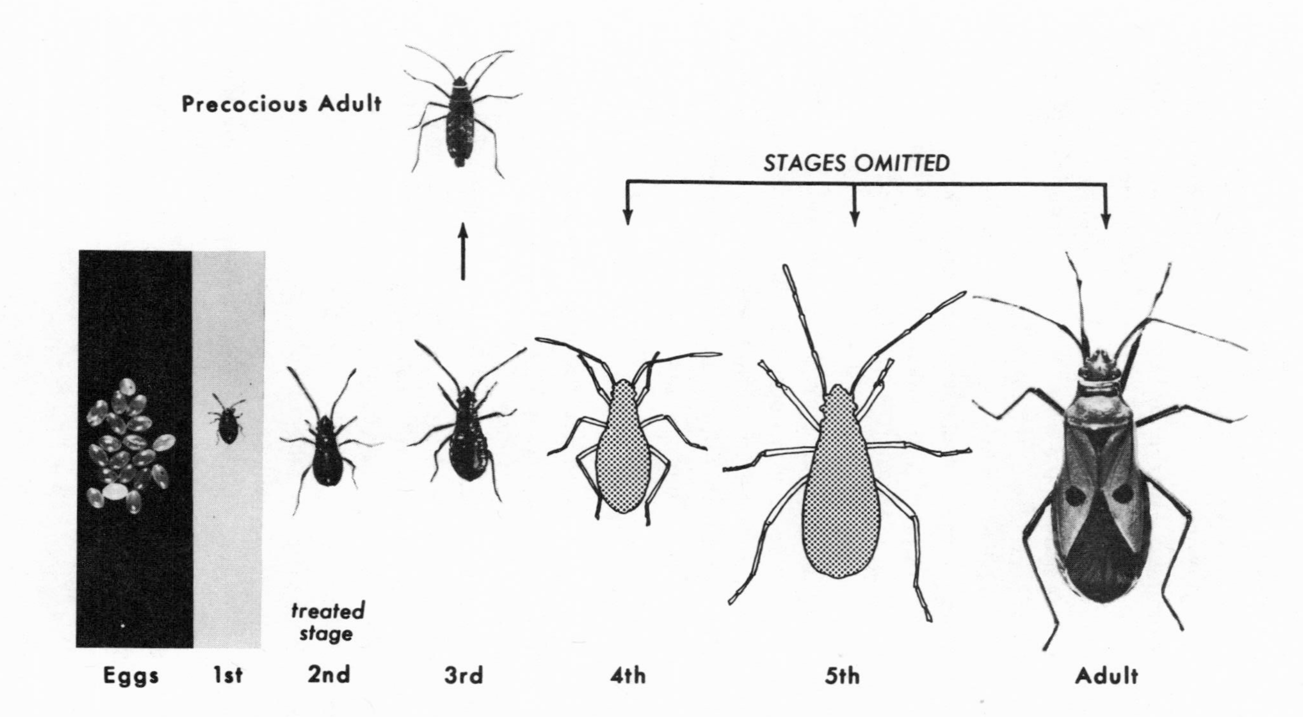

FIGURE 6. Induction of precocious, lethal metamorphosis in the cotton stainer bug by exposure to Precocene, an agent with antijuvenile hormone activity. (From W. S. Bowers *et al.*; see caption for Figure 4.)

larval insects are equipped with a brain-centered mechanism for shutting off the secretion of juvenile hormone when the larva attains a certain critical size. At the Harvard laboratory during the past few years we have busied ourselves in studying this mechanism. We have discovered yet another hormone which acts back on the brain to cause the latter to shut off the secretion of juvenile hormone by the corpora allata. Thus, indirectly it acts as an antijuvenile hormone with the potential of causing precocious, lethal metamorphosis of caterpillars and kindred larval insects. Meanwhile, in other basic studies at Cornell University, the plant *Ageratum houstonianum* (the so-called bachelor's button) was found to contain two closely related substances (see Figure 4) that apparently mimic the normal mechanism for turning off the secretion of juvenile hormone by the corpora allata. The substances have been termed Precocene I and II because of their ability to cause precocious lethal metamorphosis of certain species of larval insects (see Figure 6). Thus the stage is set for the synthesis of even more active analogs that derail the development of larval insects.

Although a new multimillion-dollar industry has come into being for making and marketing the new biorational insecticides, I do not wish to give the impression that they will necessarily prove to be a panacea, or that they promise to be a final and permanent solution to the control of insect pests and vectors of disease. Presumably, given sufficient time insects can evolve resistance to them. Manifestly, what is required is a continuation of the pure and applied research effort. Although we cannot outbreed the insects, we are well equipped to outthink them. And if we keep at it, I venture to think that the "sounds of spring" will continue to enrich our lives.

ACKNOWLEDGMENTS. The Harvard studies described in this paper were assisted by grants from the NSF, the NIH, and the Rockefeller Foundation.

Cell Biology and the Study of Behavior

ERIC R. KANDEL

My purpose in this brief discussion is to describe how cellular neurobiologists are trying to study the mechanisms of behavior and learning. For reasons that are fairly obvious, the biological study of behavior and learning is still in its infancy. Behavior is produced by the brain. The brain of man, which is the brain biologists would ultimately most like to understand, is immensely complex. In addition, because it is encased in a bony skull, it is relatively inaccessible from a technical standpoint. Even when its surface is exposed surgically by removing the bone, many important structures remain out of view because they are deeply buried within its massive substance. The brain is also inaccessible for moral reasons. Since it is the seat of those attributes and functions we refer to as character and mind, we would only want to explore the brain of man when we know exactly what we are doing, can do it safely, and have a clear medical indication for doing it. As a

ERIC R. KANDEL • Division of Neurobiology and Behavior, Departments of Physiology and Psychiatry, College of Physicians and Surgeons, Columbia University, New York, New York 10032.

result, our understanding of how the human brain controls behavior is very limited. More than in any other area of medicine, study of the relationship of brain and behavior will require the development of adequate animal models.

Fortunately, after a rather slow start, the study of how the brain controls behavior in animals other than man has advanced dramatically in the last few years. These advances have resulted from three types of development: (1) the application to the nervous system of powerful cell biology techniques for studying individual nerve cells; (2) the development of suitably simple experimental preparations, particularly certain invertebrate animals such as lobsters, snails, locusts, and leeches, in which behavior can be studied on the cellular level; and (3) the selection for study of elementary forms of behavior and learning.

The human brain and that of other higher vertebrates consists of about 10^{12} cells, a million million. The nervous systems of higher invertebrates have only ten thousand (10^4) to a hundred thousand (10^5) nerve cells. Another advantage of these nervous systems is that their cells are collected into discrete groups called *ganglia,* each of which contains only 500 to 1500 nerve cells. Despite its small number of cells, a single ganglion can sometimes control several effectors—a bodily appendage, internal organs, or glands—and can therefore mediate several behavioral responses. As a result, the number of cells committed to a *single* behavioral act may be quite small, 100 or even fewer. This numerical simplification allows one to relate the function of individual cells to behavior.

One could argue at the outset that the study of behavior is the one area in biology in which animal models, particularly invertebrate ones, are least likely to be successful. The organization of the mammalian brain and in particular the brain of man is so complex that a comparative and reductionist approach to behavior based upon a study of invertebrates is bound to fail. Man has intellectual capabilities, a highly developed language, and an ability for abstract thinking which are not found among simpler animals and which may require qualitatively different types of neuronal organization. Although these arguments are in part corrrect, they overlook certain critical issues. The question is not whether there is something special about the human brain. There clearly is. The question is, rather, whether the human brain and human behavior have *anything at all in common* with the brain and the behavior of simpler animals. If there are points of similarity, these might indicate that common principles of brain organization are involved that could profitably be studied in simple animals.

The answer to the question of similarity is clear. Detailed work by students of comparative behavior such as Konrad Lorenz, Niko Tinbergen, and Karl von Frisch have shown that man shares many common behavioral patterns with simple animals, including elementary perception, motor coordination, and simple forms of learning. These findings do not guarantee that common neuronal mechanisms are actually involved in comparable aspects of behavior or learning in invertebrates and vertebrates, but the findings are encouraging and suggest that aspects of common mechanisms are likely to be involved.

That the evolution of behavior is conservative should not be surprising. The evolution of other biological functions is similarly conservative. For example, there are no fundamental differences in structure, biochemistry, or function between the nerve cells and synapses of man and those of a snail or a leech. Since behavior is a reflection of nerve cell activity, it is perhaps not surprising that the behavior of man has features in common with that of the snail or the leech. For in behavior, as in other areas of biology, similar biological solutions are likely to be found again and again throughout phylogeny. Consequently, a complete and rigorous analysis of learning in an invertebrate, no matter how simple the animal or the task, is therefore likely to reveal a mechanism that will be of general importance. Even if a neural mechanism is not exactly identical, clearly established differences between the mechanisms of learning in different animals will enrich our understanding of learning in all animals.

This rationale has provided the logical and philosophical basis for the current research interest in the cellular mechanisms of behavior in invertebrates. The work on invertebrates, in turn, generated a number of important findings that have caused us to look at the relationship between brain and behavior in a new way.

To put into perspective the new developments that have emerged from studies of the brain and behavior of simple animals, I will briefly outline the view of the brain that existed 20 years ago, before cellular techniques were applied to the invertebrate nervous system. I will then contrast that view with the view of the brain that we now have achieved using a cellular approach.

Prior to the advent of cellular techniques, the brain was studied primarily with behavioral and surgical methods that examined the function of large groups of cells. Alternatively, the brain was studied with electrical recording and stimulating methods that examined the collective activity of large numbers of cells. Both the behavioral–surgical and the electrophysiological approaches attempted to measure physiological aspects of *whole brain function.* For example, scientists would produce lesions of the nervous system and examine how the behavior of the animal was altered by removal of a portion of its brain. This approach is evident in the pioneering work of Karl Lashley, the influential physiological psychologist who attempted to study the locus of learning in the rat by studying the effects of various brain lesions on a complex task (learning to master a maze). Lashley found that the severity of the learning defect produced by damage to the brain depended on the extent of the damage rather than the precise locus of the damage. This led Lashley and, following him, many other psychologists, to conclude that learning did not have a specific locus in the brain and therefore could not be related to specific cells. Based on these conclusions, Lashley and his contemporaries formulated a view of brain function called *mass action,* which minimized the importance of individual neurons and of specific neuronal connections. What was important to this view was brain mass, not neuronal architecture.

This view was supported by electrophysiological studies of the gross activity of the brain such as the electroencephalography or the evoked responses recorded from specific brain structures. These studies were based on the assumption that the activity of individual nerve cells was not important for behavior; what was important was the bulk or mass properties of the brain—as measured by the average response tendency of populations of nerve cells. The *mass action* or *aggregate field* view of the brain that emerged in the mid-1950s from these studies gave rise to three basic postulates of how the brain functions:

1. The various nerve cells of a brain are largely equivalent in function. The functional unit of the central nervous system is not the individual cell but an aggregate or mass of cells.
2. Behavior is generated not by specific cells but by a specific pattern of neuronal activity in a mass of central cells. The pattern is *independent* of cellular connection.
3. Learning results from changes in the statistical pattern of activity in the mass of central cells or in the surrounding chemical or electrical field.

By analyzing complex behavior with techniques that attempted to examine the brain as a whole, the aggregate field view emphasized the brain's true complexity, but it did so almost to the point of incomprehensibility. Lashley himself once remarked, after many years of study, that behavior and learning are so complex that it seemed remarkable to him that they can occur at all.

With the application to the brain of modern cell biology methods in the 1960s, the focus shifted radically from studies of the whole brain to the study of individual neurons. At first thought, this shift seemed to some students of behavior a reductionist step—almost preposterous. What can one learn about the brain, much less about those brain functions we call mind, by studying the nervous system one cell at a time?

In retrospect this reductionist step began to point the field in the right direction. It has become clear during the past 20 years that each brain, and every neural system within each brain, can best—and perhaps only—be understood in terms of the physiological and biochemical properties of its elementary units, its constituent nerve cells and their interconnections.

In the 20 years that cell biologists have been studying the brain, a new view of brain function has emerged and replaced Lashley's aggregate field view. This *cellular connection view* can also be stated in terms of three postulates:

1. The functional unit of the brain is the neuron. The various neurons of a brain are not identical and make specific connections with one another.
2. Behavior is generated by the activity of specific cells that are appropriately interconnected with other specific cells.
3. Learning results from changes in the functional effectiveness of specific cells and particularly in the specific connections between cells.

I would like briefly in this review to indicate the types of data that have been brought forward to support these basic postulates.

The functional units of the central nervous system are distinctive nerve cells. These cells make specific interconnections with one another. This postulate has two parts. One is that most neurons are not identical; many neurons are unique individuals. The other is that each neuron makes precise and invariant connections with other neurons.

Let us first consider the *invariance of neurons.* In most invertebrates that have been examined in detail the nervous system has been found to be made up of a collection of unique cells. In the nervous systems of certain animals each cell is absolutely characteristic and invariant in every member of the species. This principle, first enunciated in 1912 by Richard Goldschmidt, one of the earlier predecessors of modern cellular neurobiologists, was based on his study of the nervous system of the very simple intestinal parasite *Ascaris* (Figure 1). The brain of this worm consists of several collections of nerve cells called ganglia, separated by long nerve tracts (as shown in Figure 1A). When Goldschmidt examined the ganglia in the head of this animal, he found that they consisted of precisely 162 cells. There was never one more, never one less. This seems to hold true also for more complex invertebrates including the snail, the lobster, the crayfish, and a variety of insects that have so far been examined.

In these animals one can also study the signaling properties of neurons. One finds that although the signaling properties of all neurons have many common features, there are subtle but important differences between neurons. For example, in the abdominal ganglion of the marine snail *Aplysia,* some cells are silent and others are spontaneously active. Among those cells that are active, some cells fire regular impulses (action potentials) whereas others fire impulses in recurrent brief bursts or trains (see Figure 2).

Neurons also vary in size, position, shape, passive electrical properties, firing properties, connections, and transmitter

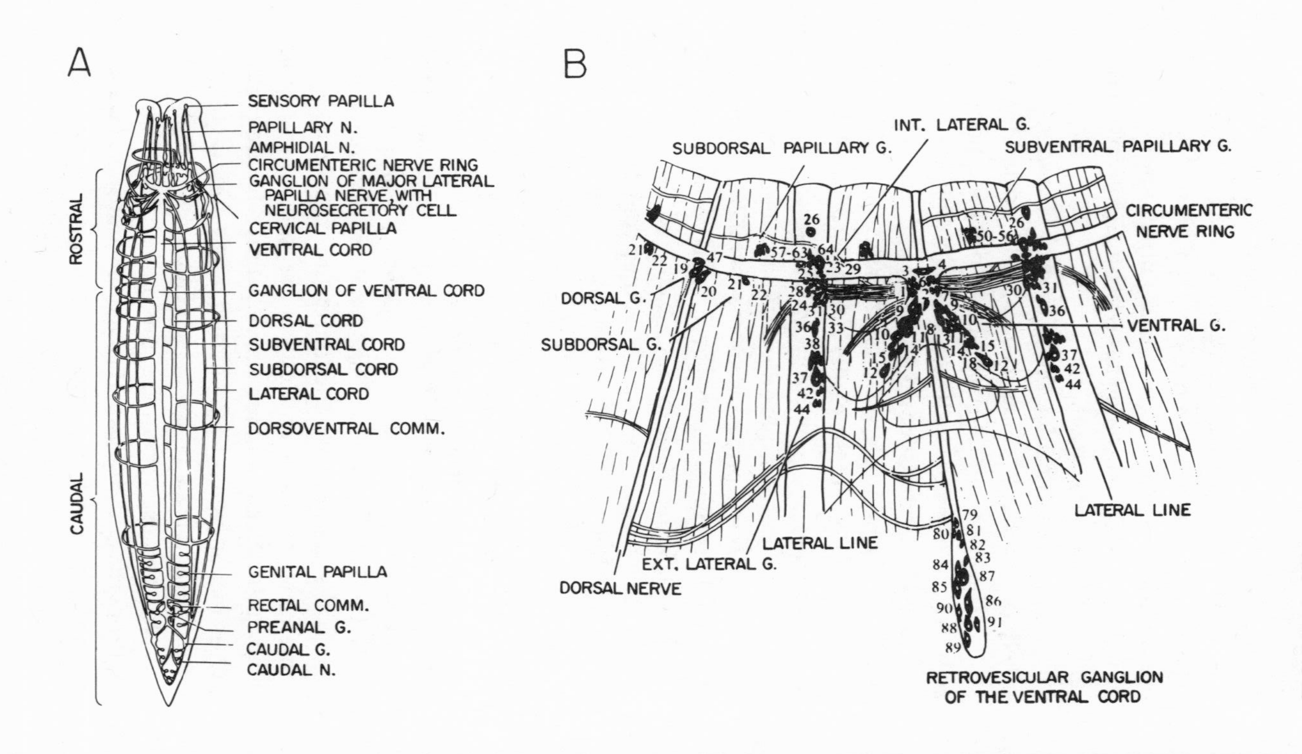
A
ROSTRAL
CAUDAL
SENSORY PAPILLA
PAPILLARY N.
AMPHIDIAL N.
CIRCUMENTERIC NERVE RING
GANGLION OF MAJOR LATERAL PAPILLA NERVE, WITH NEUROSECRETORY CELL
CERVICAL PAPILLA
VENTRAL CORD
GANGLION OF VENTRAL CORD
DORSAL CORD
SUBVENTRAL CORD
SUBDORSAL CORD
LATERAL CORD
DORSOVENTRAL COMM.
GENITAL PAPILLA
RECTAL COMM.
PREANAL G.
CAUDAL G.
CAUDAL N.
B
INT. LATERAL G.
SUBDORSAL PAPILLARY G.
SUBVENTRAL PAPILLARY G.
CIRCUMENTERIC NERVE RING
DORSAL G.
SUBDORSAL G.
VENTRAL G.
LATERAL LINE
LATERAL LINE
EXT. LATERAL G.
DORSAL NERVE
RETROVESICULAR GANGLION OF THE VENTRAL CORD

FIGURE 1. The central nervous system of the nematode worm *Ascaris lumbricoides*, illustrating the neurons identified by Goldschmidt in 1908. (Modified from Bullock and Horridge, 1965.) A: Plan of the central nervous system, illustrating the five longitudinal cords (ventral, subventral, dorsal, subdorsal, and lateral) and the horizontal commissures. The nervous system can be roughly divided into a rostral portion, containing 162 cells, and a caudal portion, containing about 40 cells. B: Rostral portion of the nervous system, illustrating the identified cells. The esophagus and lips have been removed and the cylindrical body wall has been incised and unrolled. Most of the identified cells are marked with their identifying number.

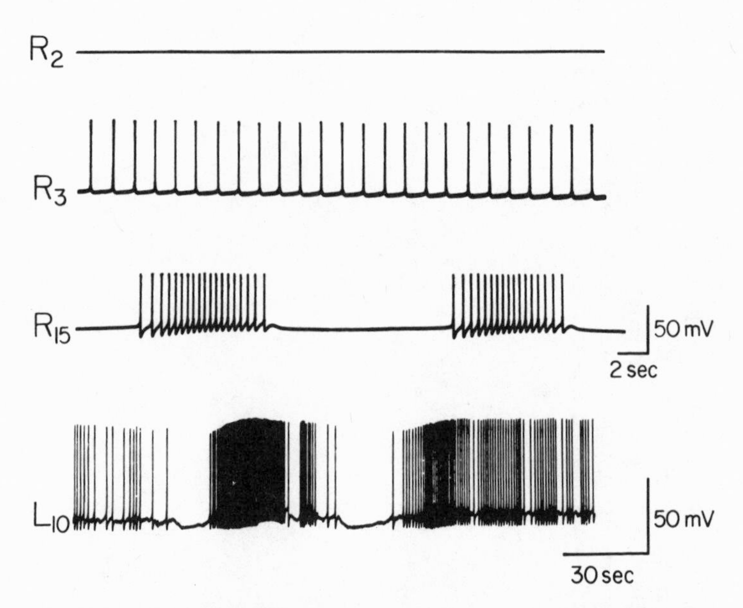

FIGURE 2. Intracellular recordings showing four types of firing patterns from cells in the abdominal ganglion of the marine snail *Aplysia:* R_2, silent; R_3, regular beating rhythm; R_{15}, regular bursting rhythm; L_{10}, irregular bursting rhythm. (Time and voltage calibrations for L_{10} are different from those for the other cells.) (From Kandel, 1976.)

biochemistry. Whether or not each cell in the mammalian nervous system is also a distinct individual is not yet known. However, recent studies in the sensory systems of mammals by Hubel and Wiesel at the Harvard Medical School and by Mountcastle and his colleagues at the Johns Hopkins Medical School have revealed fascinating and important differences between neighboring neurons. Independent studies of the development of the vertebrate brain by Roger Sperry at the California Institute of Technology lead to a similar conclusion.

The connections between cells are precise and invariant. Cells always make the same kinds of connections to other cells. This applies not only for the *presence* of connections but also holds for the *sign* of the connections (that is, whether the connections are inhibitory or excitatory). For example, there are some cells in invertebrates that mediate different actions through their various connections. Figure 3 illustrates one of these cells, located in the abdominal ganglion of *Aplysia.* This cell excites some follower cells, inhibits others. It also makes a dual (excitatory–inhibitory) connection to one cell, which is not illustrated in this figure.

Figure 4 is a map of the connections made by this one cell. The cell always connects to exactly the same cells, in exactly the same way. It always excites precisely the same cells, always inhibits the same cells, and always makes a dual connection to the same cell.

Behavior is generated by the activity of appropriately interconnected cells. Since individual cells connect invariably to the same follower cells and can mediate actions of different sign, a cell located at a critical point in the nervous system is in a position to control a discrete behavioral act. This principle was first enunciated by C. A. G. Wiersma working at the California Institute of Technology. Wiersma called these critically situated neurons "command cells" because the individual cell could command a whole behavioral sequence. Command cells have now been found in all higher invertebrates so far examined. This is illustrated in Figure 5, which shows the connections to the heart of *Aplysia* made by the multiaction cell illustrated in Figures 3 and 4. Activity of this one cell increases heart rate and cardiac output. It does so by exciting the cells that speed up the heart while inhibiting the cells that inhibit the heart, and also by inhibiting the cells that constrict the major blood vessels. As a result of the actions of this one cell, the heart beats more rapidly and pumps more blood to the tissues.

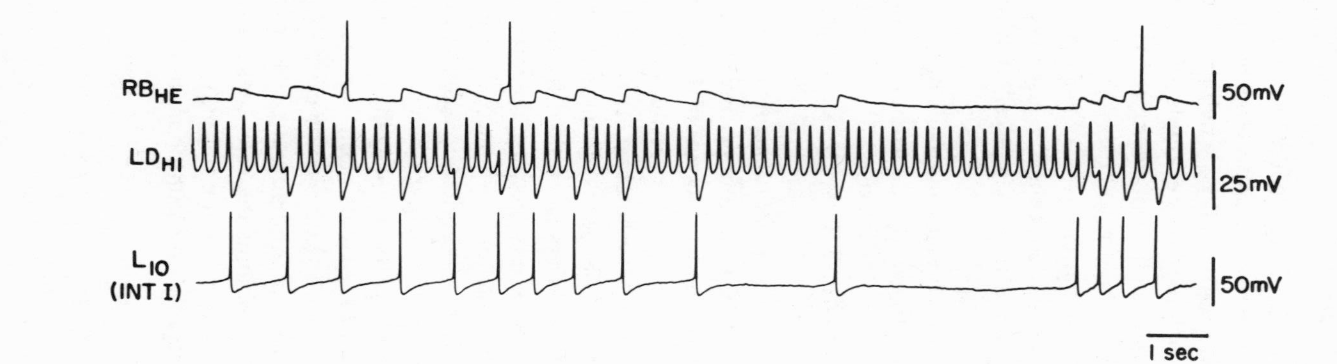

FIGURE 3. Identified cell L_{10} (Interneuron I) produces opposite synaptic action on heart motor neurons in *Aplysia*. Simultaneous intracellular recordings from all three cells are shown. Each impulse in L_{10} excites RB_{HE} and inhibits LD_{HI}. (From Koester *et al.*, 1974.)

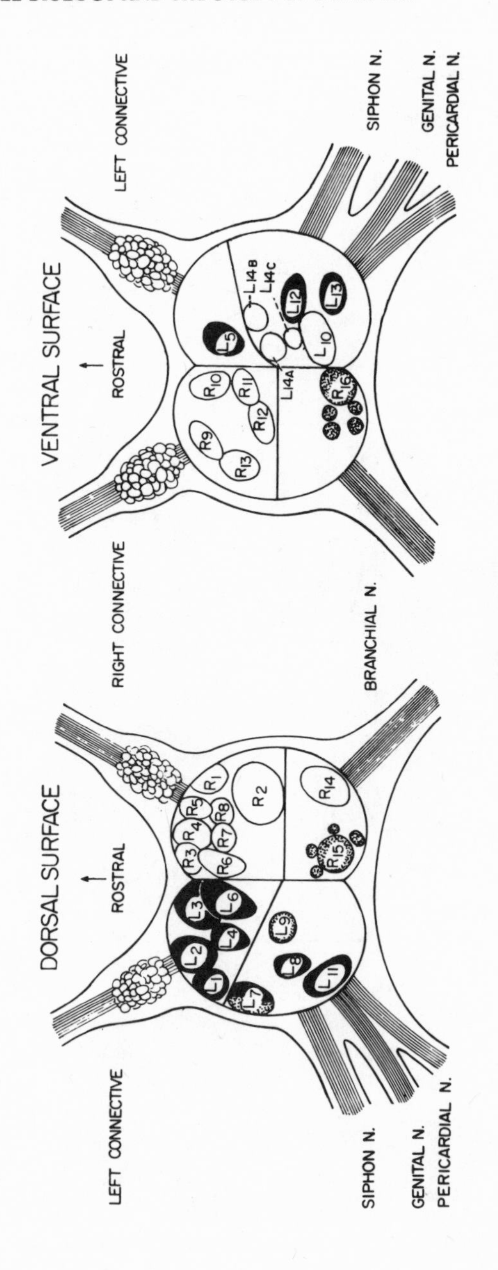

FIGURE 4. Idealized drawing of the dorsal and ventral surfaces of the abdominal ganglion of *Aplysia* illustrating the position of a number of unique cells called "identified" cells. Many of the cells receive invariant inhibitory connections from L_{10} (indicated in black) and other cells receive excitatory connections (stippled). One cell receives a dual excitatory–inhibitory connection (black and stippled). (From Kandel *et al.*, 1967.)

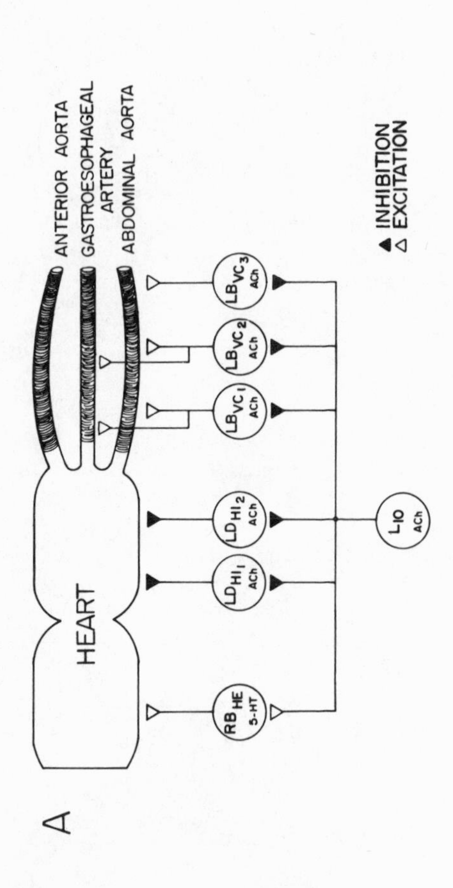
A
HEART
ANTERIOR AORTA
GASTROESOPHAGEAL ARTERY
ABDOMINAL AORTA
RB HE 5-HT
LD HI1 ACh
LD HI2 ACh
LB VC1 ACh
LB VC2 ACh
LB VC3 ACh
L10 ACh
▲ INHIBITION
△ EXCITATION

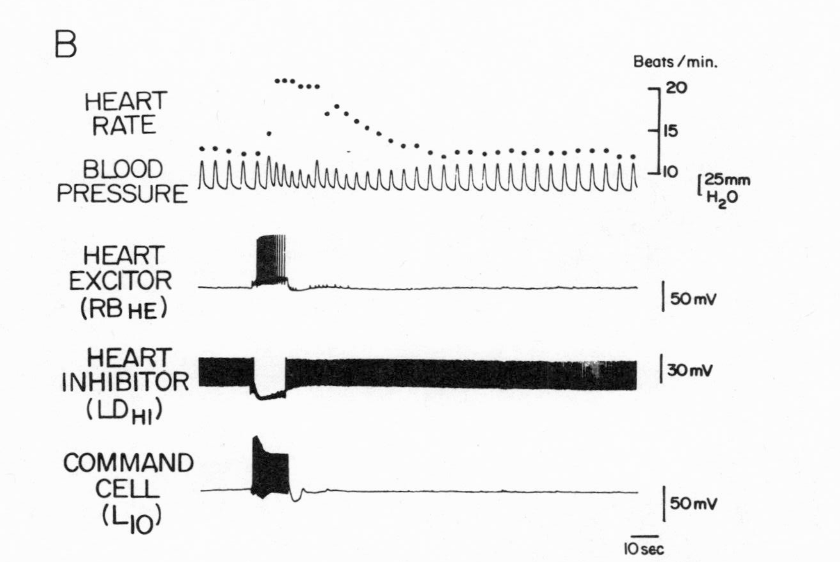

FIGURE 5. A behavioral act triggered by a command cell. A: Diagram of the synaptic connections made by command cell L_{10} to the cardiovascular motor cells. Where known, the transmitter utilized by each cell is indicated. ACh refers to acetylcholine, 5-HT to serotonin. (From Koester *et al.*, 1974.) B: A behavioral act initiated by the activity of one cell, the command cell L_{10}. A brief burst of impulses in L_{10} excites the heart excitor motor neuron (RB_{HE}) and inhibits the heart inhibitor (LD_{HI}). This is associated with a long-lasting increase in heart rate and a short increase, followed by a decrease, in blood pressure. (From Koester *et al.*, 1974.)

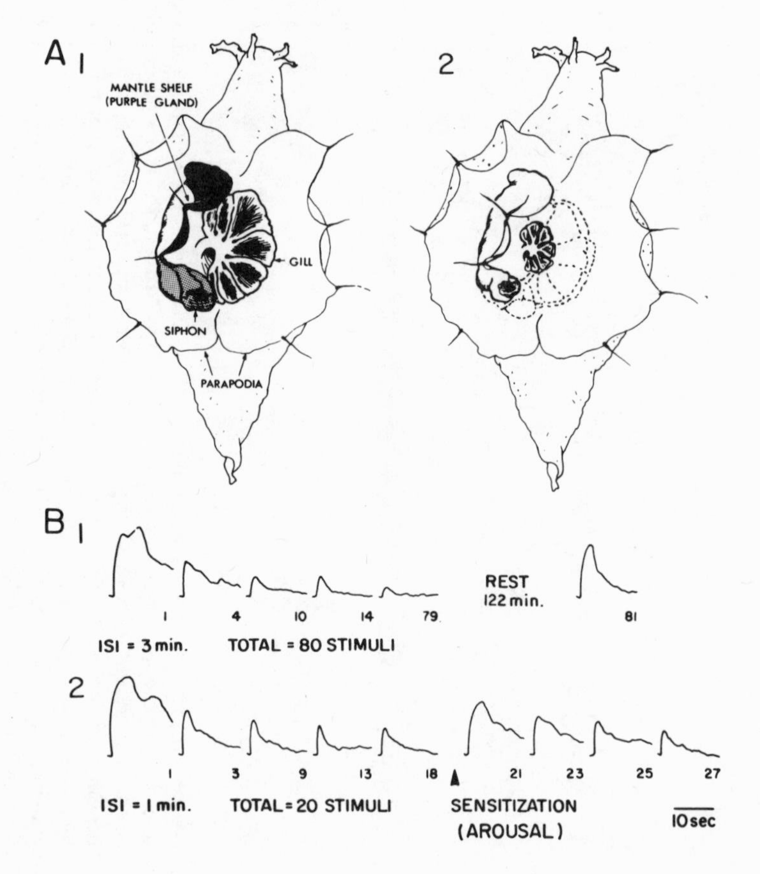
A 1
MANTLE SHELF
(PURPLE GLAND)
GILL
SIPHON
PARAPODIA
2
B 1
REST
122 min.
1
4
10
14
79
81
ISI = 3 min.
TOTAL = 80 STIMULI
2
1
3
9
13
18
21
23
25
27
ISI = 1 min.
TOTAL = 20 STIMULI
SENSITIZATION
(AROUSAL)
10 sec

←

FIGURE 6. Two simple forms of learning: habituation and sensitization (arousal). A: The defensive withdrawal reflex of the siphon and gill in *Aplysia* (dorsal view of an intact animal). Since the parapodia and mantle shelf ordinarily obscure the view of the gill, they must be retracted to allow direct observation. The tactile receptive field for the gill withdrawal reflex consists of the siphon and the edge of the mantle shelf. A_1: relaxed position; A_2: defensive withdrawal reflex in response to a weak tactile stimulus to the siphon (a jet of seawater). The relaxed position of the gill is indicated by the dotted lines. (From Kandel, 1976.) B: Short-term habituation and dishabituation of the gill withdrawal reflex in *Aplysia*. Habituation, spontaneous recovery, and dishabituation of the gill withdrawal reflex. Photocell records of gill withdrawal from two training sessions with the same animal. B_1: Habituation of the response with 80 repetitions of the stimulus at 3-min intervals. The major decrease occurs within the first 10 stimuli. Following a 122-min rest, the response recovered partially. B_2: A later training session after the animal had fully recovered. The response was habituated a second time at 1-min intervals. A sensitizing stimulus, consisting of a strong and prolonged tactile stimulus to the neck region, was presented. Following the sensitizing stimulus, responses were facilitated for several minutes. (From Pinsker *et al.*, 1970.)

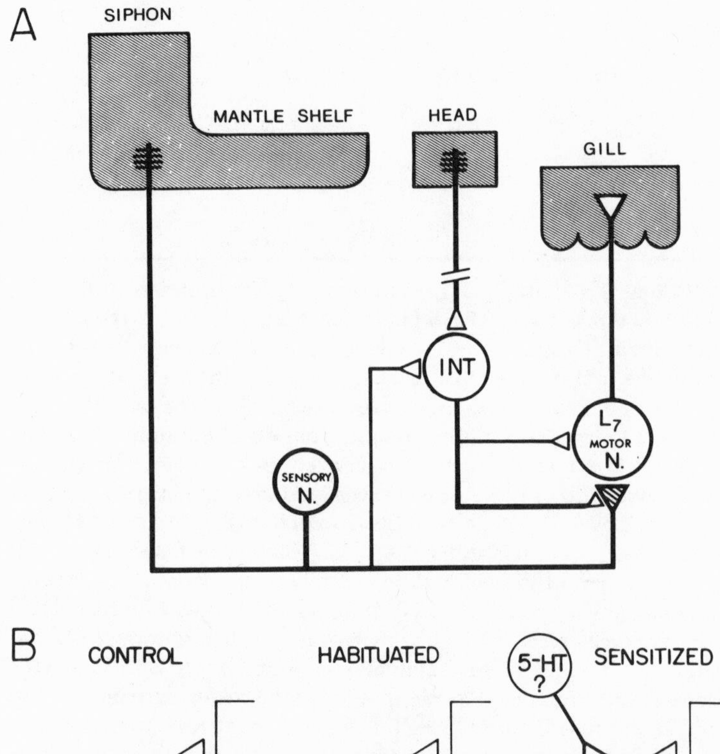

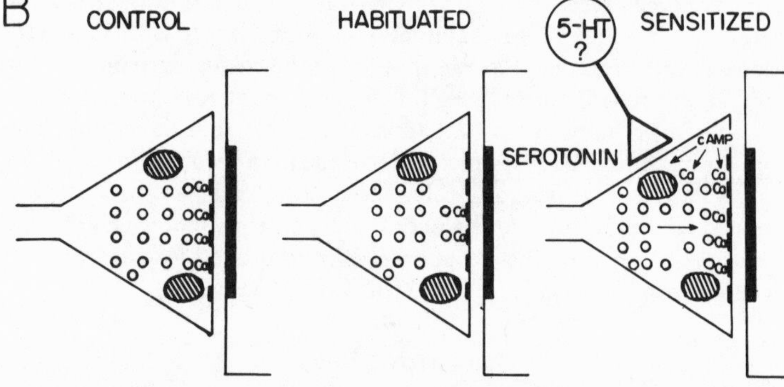

FIGURE 7. Cellular mechanisms of habituation and sensitization. A. Postulated plastic changes in the circuit underlying habituation and sensitization of the gill-withdrawal reflex. For schematic purposes only one sensory neuron and one motor neuron (L_7) are illustrated. Similar processes are thought to be operative at all the synaptic

This is only a simple example of a command cell. In the crayfish, and even in a much more complex animal like the goldfish, a single impulse in a single command neuron causes a complete escape response. Recently, Mountcastle has suggested that small groups of cells may serve similar command functions in the primate brain to control purposeful voluntary movements.

Learning involves a change in the functional effectiveness of specific cells and their connections. The finding that behavior is mediated by invariant cells that interconnect in precise and invariant ways seems at first to pose a paradox for the study of learning. How is behavior modified in a precisely wired neural circuit? The few studies that have so far been carried out on

---

contacts made by the population of sensory neurons on the other motor neurons and command cells. The terminal of a sensory neuron is the common locus of both the synaptic depression underlying habituation and the presynaptic facilitation underlying sensitization. The pathway from the head mediating presynaptic facilitation ends on an excitatory command cell (L22 or L23) that synapses on the sensory neuron terminal. The plastic synapse from the sensory to motor neuron is indicated in gray. (From Kandel *et al.*, 1976.) B. Suggested molecular mechanisms for depression and facilitation based on vesicle mobilization. Each consecutive impulse in the terminal of a sensory neuron may lead to a progressively smaller increase in free Ca^{2+} either because of the high density of mitochondria that take up free Ca^{2+} as it comes in with the impulse or because of a progressive decrease in the Ca^{2+} permeability of the terminal membrane. As a result, repeated stimulation does not lead to effective mobilization of vesicles to the release sites at the terminal, and a partial depletion of vesicles is achieved after 5–15 impulses. This could account for habituation. Sensitization is mediated by command cells that utilize serotonin, and it produces two complementary actions: (1) a synaptic action, which increases the Ca^{2+} concentration and briefly enhances mobilization, and (2) a prolonged increase in cyclic AMP, which increases Ca^{2+} levels by inhibiting Ca^{2+} uptake by the mitochondria or by increasing Ca^{2+} influx, thereby enhancing mobilization of transmitter vesicles. (From Kandel, 1976.)

learning have shown that this paradox has a rather simple answer. At least in its most elementary forms, learning seems to involve a change in the functional effectiveness of previously existing conditions. Thus, whereas genetic and developmental processes determine the functional properties of neurons as well as the precise interconnections, the genetic and developmental programs do not specify fully the strength of the connections between certain cells. The long-term strength of these connections is determined by environmental factors, such as learning.

This conclusion comes in part from the study of very simple forms of learning such as habituation and sensitization. Habituation is the process whereby an animal learns to ignore a stimulus that becomes trivial with repetition. When an animal is first exposed to a new stimulus, it gives a very brisk response. With repeated stimulation the animal habituates to the stimulus and the response becomes progressively weaker and weaker. The animal will now retain a memory of that stimulus (it will remember to ignore it) for periods ranging from minutes to weeks, depending on the extent or history of exposure to the stimulus. However, if now a noxious stimulus is presented, the animal will become aroused and will show sensitization: it will instantaneously give a full response to the stimulus which it previously had ignored (Figure 6).

Habituation and sensitization are common learning experiences in man. They also can be demonstrated in simple defensive reflex behaviors of invertebrates such as the defensive withdrawal of the gill, a respiratory appendage, in the marine snail *Aplysia* (see Figure 6). Cellular analyses of habituation and sensitization of this reflex show that these two forms of learning involve an alteration in the functional effectiveness of a specific set of connections between the sensory and motor neurons (Figure 7A). Habituation results from a progressive decrease in the effectiveness of this particular set of connectives. Sensitization leads to a sudden restoration of effectiveness.

Recently it has become possible to develop a preliminary molecular model of habituation and sensitization based on evidence from a number of different experiments (Figure 7). According to this model the repeated activity that accompanies habituation leads to a decrease in the amount of transmitter released from the connections made by the sensory neurons. Sensitization leads to a sudden increase in the amount of transmitter substance at the same set of synaptic connections. Enhancement of transmitter release is produced by several (arousal) neurons that act directly on the terminals of the sensory neuron and modulate the ability to release the synaptic transmitter substance. There is some evidence that the facilitating neurons use the substance serotonin as their chemical mediator. Serotonin is postulated to act directly on the terminals of the sensory neurons to increase the concentrations of an intracellular messenger called cyclic AMP. Cyclic AMP, in turn, is thought to increase the free calcium concentration in the terminals. The increase in free calcium leads directly to an enhancement of transmitter release.

The availability of simple systems such as the one I have outlined here should make it possible to explore a wide range of questions that have traditionally interested students of behavior. For example, there has been persistent concern about the relationship of short-term memory and long-term memory. Is long-term memory a process distinct from short-term memory, or do both types of memory involve variants of a single process, a single memory trace? It may be that short-term habituation involves a transient depletion of only a particularly accessible fraction of the transmitter pool of the sensory neuron terminals, a readily releasable pool of transmitter. Long-term habituation seems to involve a prolonged depression of the synthesis of transmitter and of the transport of the transmitter vesicles from the cell body to the terminals. Studies in simple systems should be able to test hypotheses such as these directly.

Finally, cellular analysis can also be applied to behavioral abnormalities. This represents, I think, one of the most exciting areas for clinical psychiatry. Although it has long been recognized that some behavioral disturbances originate as excessive expressions of normal behavior, few attempts have been made to develop simple animal models of abnormal behavior that would allow a cellular analysis of how normal behavior can be distorted by experiential or genetic factors. It is unlikely that invertebrates will yield perfect models of abnormal behavior that will include all essential components of human abnormalities. However, these models may well illustrate some of the general mechanisms by which environmental factors exaggerate or disrupt normal adaptive responses.

For example, anxiety or fear is a normal response to threat, whether to one's person, attitudes, or self-esteem. Normal anxiety is adaptive: it signals potential danger and can contribute to the mastery of a difficult situation, and thus to personal growth. Excessive fear responses are, on the other hand, maladaptive and are commonly seen in neurotic and psychotic conditions. However, excessive fear responses (extreme forms of sensitization) are also seen in invertebrates and their analysis could provide new insight into this behavioral abnormality.

I have here considered only studies that combine cellular biological and behavioral approaches. However, much is also being learned by combining behavioral and genetic techniques as has been done by Seymour Benzer at the California Institute of Technology. Benzer (1973) used a variety of mutants of the fly *Drosophila* to study behavioral abnormalities, including visual disturbances, abnormalities in circadian rhythm and sexual behavior, muscular dystrophies, and sudden cessation of development. This approach will certainly shed much light on the relation of genetic to experiential factors in the control of behavior.

In summary, I would not like to give you the impression that we have come a long way in understanding the mechanisms of behavior and learning; the road is long and we are at the very beginning. I would like, however, to leave you with the idea that we are on the right road and that cellular biological approaches can now be applied effectively to the study of behavior. As a result, we hopefully will be in a position in the future to provide increasingly more satisfactory explanations of progressively more complex instances of behavior, learning, and their abnormalities.

ACKNOWLEDGMENTS. I have benefited from the comments on this paper by J. H. Schwartz, W. A. Spencer, and Sally Muir. I am grateful to Kathrin Hilten for preparing the figures. The original research was supported by NIMH Career Scientist Award 5–KO5–MH–18558–09; NIMH grant RO1–MH–26212 –02; and NINCDS grants 2–RO1–NS–12744–02 and GM–23540–01.

REFERENCES

Bentley, D., and Hoy, R. R., 1974, The neurobiology of cricket song, *Sci. Am.* **231**:34.

Benzer, S., 1973, Genetic dissection of behavior, *Sci. Am.* **229**:24.

Bullock, T. H., and Horridge, G. A., 1965, *Structure and Function in the Nervous Systems of Invertebrates* (2 vols.), W. H. Freeman, San Francisco.

Kandel, E. R., Frazier, W. T., Waziri, R., and Coggeshall, R. E., 1967, Direct and common connections among identified neurons in *Aplysia*, *J. Neurophysiol.* **30**:1352.

Kandel, E. R., 1976, *Cellular Basis of Behavior: An Introduction to Behavioral Neurobiology*, W. H. Freeman, San Francisco.

Kandel, E. R., Brunelli, M., Byrne, J., and Castellucci, V., 1976, A common presynaptic locus for the synaptic changes underlying short-term habituation and sensitization of the gill-withdrawal reflex in *Aplysia*, *Cold Spring Harbor Laboratory Symposium on Quantitative Biology, XL: The Synapse*, pp. 465–482.

Koester, J., Mayeri, E., Liebeswar, G., and Kandel, E. R., 1974, Neural control of circulation in *Aplysia*. II. Interneurons, *J. Neurophysiol.* **37**:476.

Nicholls, J. G., and Van Essen, D., 1974. The nervous system of the leech, *Sci. Am.* **230**(1):38.

Pinsker, H., Kupfermann, I., Castellucci, V., and Kandel, E. R., 1970, Habituation and dishabituation of the gill-withdrawal reflex in *Aplysia*, *Science* **167**:1740.

The Introduction of Missing Enzymes into Human Cells

Alternatives to DNA Recombinancy in Genetic Disease

GERALD WEISSMANN

Pendant que les fonds publics s'écoulent en fêtes de fraternité, il sonne une cloche de feu rose dans les nuages.

—Rambaud

The hazards and promises of new research have probably never been as carefully anticipated as those arising from studies of DNA recombinance. It is to the solid credit of the discoverers of gene splicing that they themselves have warned us of its possible risks, have convened meetings to discuss the social consequences of its applications, and have been instrumental in elaborating what they perceive to be appropriate

GERALD WEISSMANN • Professor of Medicine and Director, Division of Rheumatology, New York University School of Medicine, New York, New York 10016.

guidelines for its safe exploitation. Yet even this responsible, cautious approach has not prevented the emergence of frequently harsh (and occasionally cogent) criticism both from within and without the scientific community. Having gone this far, the critics say, we should perhaps go no further into the business of mucking around with the genomes of man or microbes until we've thought a lot more about the whole problem.

Now it will not be my business directly to rush where the angels of Cambridge or Palo Alto have already trodden (*The Sciences* of October 1976 has already covered some of this territory). Rather, I would like to suggest that alternatives to DNA splicing are available in at least one area where genomic intervention has been proposed: the repair of genetically determined enzyme deficiencies of man.

A policy statement by the NIH (*Federal Register*, July 7, 1976) outlines the most conservative prospects for the immediate applications of successful experiments with DNA recombinance:

> Within the past decade, enzymes capable of breaking DNA strands at specific sites and of compiling the broken fragments in new combinations were discovered, thus making possible the insertion of foreign genes into viruses or certain cell particles (plasmids). These in turn, can be used as vectors to introduce the foreign genes into bacteria or into cells of plants or animals in test tubes. Thus transplanted, the genes may impart their hereditary properties to new hosts. These cells can be isolated and cloned—that is, bred into a genetically homogeneous culture. In general, there are two potential uses for the clones so produced: as a tool for studying the transferred genes and as a new useful agent, say for the production of a scarce hormone.

But the very real prospect of using such vectors to repair human genetic deficiency *in vitro* and *in vivo* is also before us. If we can begin to repair, in the dish, cultured marrow cells of a patient with sickle cell disease, or the pancreatic cells of a patient with diabetes, then surely only problems related to delivery (engineering) separate us from the goal of substituting good genes for bad in the clinic. With this bait of therapy dangling before them, the carp who dwell in pools of public funding should seize the line. What's the hook? I would suggest that the delivery problem is not simple, that its solution will consume considerable time and money, and that its dangers may be of greater consequence than the *Andromeda Strain* fantasies which seem to have motivated the City Council of Cambridge.

Now the insertion of normal genes into afflicted humans will have to follow *some* of the strategies employed by viruses: those pathogens which preempt our cellular manufactories for their own procreation. I have little doubt that given luck, money, and time, we shall be able to mimic these strategies for the purpose of splicing good genes into appropriate areas of human DNA. By suitable engineering we may eventually be able to "catch" a reparative gene as readily as the flu.

But the hook is error! We really do not yet understand all the "on–off" signals of gene action in living cells: why, for example, our bone cells do not make fingernails or skin. One might therefore anticipate either disastrous or amusing results—depending on one's sense of morality—in the early trials. Unfortunately, once we have successfully spliced reparative genes into the strands of normal DNA, we cannot recall them for countless generations, and each feasible error in the laboratory or clinic may grow up to haunt or govern us. Bad enough to think of skin cells making hemoglobin, a bloody nuisance, but imagine the behavioral consequences of aberrant gene products released by brain cells! Now it will probably be correctly reasoned that if we become clever enough to insinuate purified genes into the serpentine coils of mammalian chromosomes, we will, by virtue of the experience itself, have become clever enough to avoid such picturesque disasters. So be it. But let us see what technology is available for more immediate exploitation before we decide if the whole enterprise of DNA splicing should account for a major fraction of our effort in genetic disease.

Genetic human diseases are usually characterized by diminished biological activity of specific ferments (enzymes). Whereas this abnormality frequently disables those unlucky enough to bear a *double* dose of the aberrant gene (homozygotes), it may sometimes confer a real or potential advantage on carriers of a *single* aberrant gene (heterozygotes). Consequently, "eugenic" interventions designed to assure the disappearance of the bad gene by controlled mating may be not only socially presumptuous but biologically ill-advised. It therefore follows that if we cannot prevent their transmission, we should at least treat the consequence of aberrant genes lurking in our midst.

The abnormal or absent gene products may be arbitrarily divided into four groups.

Abnormal proteins may be elaborated with *functional* nonenzymatic properties which differ from the normal. Thus, the amino acid substitutions in sickle cell hemoglobin lead to structural abnormalities of affected red cells.

In other circumstances, as in some forms of diabetes, or of fat accumulation in the blood, *structural* components of the cell membrane (receptors) are absent and functional abnormalities such as high blood levels of fat or sugar result.

In yet other diseases, *cytoskeletal* components may be absent, and proteins which help cells move or retain their shape (microtubules) are inadequately linked in the genetic disease of Kartagener's syndrome. Such patients cannot cleanse their airways, and worse, their sperm cannot move properly.

Finally, and perhaps most prevalently, the missing protein is an *enzyme* (we use the suffix "-ase" to describe these proteins) required for vital body functions. Lack of enzyme activity leads both to bodily deficiencies of enzyme product (output) and to excess of its substrate (input). An analogous lack of "Aswan-ase" in Egypt would lead to flooding of Lake Victoria and drought in the Nile delta.

It is this latter group, the inheritable enzyme deficiencies of man, that has occupied the greatest attention. At least three score such deficiencies have been well identified in man and of these, over 30 have been found to be due to loss of a specific enzyme ordinarily sequestered within lysosomes of the cell (Hirschhorn and Weissmann, 1977). Lysosomes are submicroscopic sacs within most cells that are surrounded by thin lipid membranes. These containers are filled with enzymes designed to degrade material taken in from outside the cell or worn-out molecules which accumulate within the cell. In a sense, lysosomes constitute the internal digestive tract of cells. Therefore, when such an enzyme is missing, both externally and internally derived residues accumulate and, as in the digestive tract of man, constipation results.

This form of "cellular constipation" with fats, sugars, or proteinaceous material results in significant injury to the cell and tissue. Many human diseases, such as Tay–Sachs disease, Gaucher's disease, and the Hunter–Hurler syndrome are due to such specific defects of the intracellular digestive apparatus. In these diseases, which may appear in early childhood and sometimes lead to early death, the accumulation of digestive residues in lysosomes of vital organs such as the liver or spleen causes the organs to become markedly enlarged. Occasionally, even the nervous system is involved and the fatty building blocks of nervous tissue accumulate until functions of the brain are impaired. This frightening group of diseases, defined in considerable part through the work of Hers in Louvain, Neufeld and Brady in Bethesda, and Dorfman in Chicago, has been attacked most vigorously in laboratory and clinic (Hirschhorn and Weissmann, 1977).

However, another group of enzyme deficiencies also exist in which the enzymes are missing not from within membrane-bounded organelles but from the very sap of the cell itself, the cytosol. Some of these diseases include a very disabling deficiency of the enzyme adenosine deaminase. Absence or malfunction of this enzyme renders small children unable to mount proper immune responses. Such children, who suffer from the condition of "severe combined immunodeficiency," are condemned at the present time to spend most of their life in germ-free, isolated, plastic environments. Their plight, due directly to the absence of enzymes required for the normal maturation of the immune cells, is indirectly related to the accumulation of substrate. Another such disease, known as the Lesch–Nyhan syndrome, is associated with self-mutilation and central nervous system damage. The enzyme missing is also related to the synthesis of purines, which leads to manifestations in innocent childhood of what historically had been considered a disease of adult dissipation: gout.

It is in the group of inheritable enzyme deficiencies that the most progress has been made toward reconstituting affected individuals. Indeed, several distinct, promising paths have already been explored. First, it was necessary, for each of these diseases, to define the stored substrate (input) as well as to identify and isolate the normal enzyme which is responsible for its cleavage (output). It was next necessary to demonstrate the absence or malformation of the enzyme in the affected individuals and to outline the mode of inheritance. This task has to a large degree been accomplished. Indeed, many of these enzymes have been purified to near homogeneity from human sources, such as placenta, white blood cells, etc. Now, once the enzyme is isolated the problem evolves itself into the engineering feat of delivering the enzyme to the tissues from which it is missing. Far less progress has been made toward this goal, but some results are in already; every month brings news of effort and partial success.

The most direct enterprise, due in large measure to Roscoe Brady and his co-workers (Brady *et al.*, 1973, 1974) in Bethesda, has involved injecting large quantities of purified enzymes directly into the bloodstream of affected individuals. In several cases of Gaucher's and related diseases, injections of the purified enzyme were shown to induce a breakdown of accumulated substrates in the blood, and to a lesser degree in the vital organs. This work demonstrated that it is indeed possible to deliver an enzyme which can effect removal of the offending material as if the enzyme had been manufactured by the deficient individual himself. The difficulties with this approach are several. In the first place, very little of the actual enzyme which is infused winds up reaching the organs that are affected, especially the central nervous system. Second, were the material to be injected repeatedly it is by no means clear that the individual would not mount an immune response to the injected protein, which at the least is somewhat foreign.

Another approach which has been attempted is to introduce into the affected patient cells or even whole organs that can either carry or manufacture the missing enzyme. Thus, for example, it is possible to trap the missing enzyme inside reformed, compatible human red cells. It has been known for some time that it is possible to prepare red cell "ghosts." Red cell ghosts resemble the skins of grapes which have been pitted and had the pulp extracted. The emptied "skin" of the red cell grape tends to seal itself (a resealed ghost), and after proper manipulation in the laboratory the ghosts can be persuaded to trap purified enzymes which the host is missing. If the red cells are compatible, the recipient can now accept enzyme-laden ghosts without markedly untoward consequences. Consequently, the enzyme would be expected to be liberated at those sites in the recipient wherein red cells are normally turned over. This technique offers significant hope for reconstitution of enzyme deficiencies involving those organs which ordinarily break down erythrocytes. The technique, however, is cumbersome, difficult, and as yet not too efficient.

In yet other, but related trials, clinicians have transplanted whole organs, such as kidneys, from compatible normal donors to deficient patients. The transplanted organ liberates the enzyme (of which it elaborates a normal amount) into fluids of the deficient host. As in infusions of exogenous, purified enzymes, it is assumed that these small quanta of enzymic activity released into the circulation will find their way to various parts of the body. The difficulties here are that the amounts of enzyme made by the tissue *in vivo* are small, that very little of the exogenous enzyme tends to enter other tissues, and finally, that the problem of reaching all affected organs has by no means been solved.

Another approach is based on the discovery that forms of enzymes possessing certain sugar groupings (sialic acid) at critical points in the molecule are taken up more avidly by some cells or organs. By appropriately modifying the sugar residues of purified enzymes it may be possible selectively to trick cells into recognizing that these "high-uptake forms" are designed to be introjected into the affected tissues. This technique has yielded encouraging results in tissue culture but has not yet been tested appropriately in human disease.

The last of these enterprises was launched in our laboratory (Sessa and Weissmann, 1970) and in the laboratories of G. Gregoriadis and B. Ryman (Gregoriadis and Ryman, 1972; Gregoriadis *et al.*, 1971) in London. It is based on the fortuitous confluence of two streams of investigation.

The first stream of investigation was the one outlined above: elucidation at the cellular and biochemical level of the lysosomal storage diseases in which the intracellular digestive tract of cells lacks specific enzymes. The parallel stream emerged from other studies, by no means directed toward genetic storage diseases. Research carried out with A. D.

Bangham in Cambridge (Bangham *et al.*, 1965) and in our own laboratory (Sessa and Weissmann, 1968) showed that it was possible to create models of cellular membranes in the test tube. One of the general approaches of scientists to the study of living systems is to construct with artificial, defined materials a reasonable working model of that system. Indeed, we found that just as the natural membranes of lysosomes enclose enzymes, segregating these from their surroundings, so also can membranes produced in the laboratory be persuaded to sequester purified enzymes from salty solutions of water in the test tube. In techniques now simplified enough to be repeated in any college chemistry laboratory, the same sort of fatty materials which have been found in biological membranes are dissolved in chloroform. After the chloroform is evaporated, the fats form a thin layer at the bottom of the reaction flask. Then, after addition of enzyme-containing solution, the fatty substances leave the surface of the flask and spontaneously form into small vesicles, the size of which can be controlled by the experimenter. Such vesicles are filled both with the enzyme desired to be trapped and with the watery solution in which it was dissolved. Separation by means of chromatographic techniques makes it possible to purify these enzyme-containing fatty vesicles, which we have called "liposomes" (Sessa and Weissmann, 1968). We now have in hand a kind of rudimentary "artificial lysosome."

Indeed, this process of sequestering an enzyme from watery solutions containing salts and other dissolved materials may very well resemble events at the very beginning of living things. Were lipids, of whatever origin, to have found themselves in the vicinity of primordial reactions involving DNA, RNA, and protein synthesis, it would not be too difficult to imagine them forming self-assembled bubbles within which to segregate this new thing—life, as it were—from the hostile sea.

Well, the experimental outcome of enzyme trapping in

liposomes was not only an exercise in the building of models, but unexpected dividends resulted in the utility of these models for intervening in genetic diseases of man. Studies of genetic diseases and liposomes came together in what I like to refer to as the "Trojan horse" approach to genetic engineering. Most cells will not take up such lipid vesicles very readily. Nor, as we have discussed above, will they take up the purified enzymes. This, as in diplomatic circles, may be due to failure of recognition: is the material friend or foe? However, various cells *can* recognize (by means of surface receptors) those substances with which it must exchange physiologic embassy.

The most basic system for the recognition of foreignness in mammals is the immune system. Thus, scavenger cells tend to recognize foreign invaders when the immune system of the host has tarred them with the brush of antibody. Realizing that this immune recognition system might serve to engage cell receptors of cells from genetically deficient hosts, we coated our liposomes with antibodies in the test tube. We were gratified to find that living cells which display receptors for these antibodies (leukocytes) were persuaded to ingest enzyme-containing liposomes as if these laboratory artifacts were bacteria or viruses against which the host had mounted an immune response. Soon, phagocytic cells, such as those of liver, spleen, bone marrow, or blood, were capable, at least *in vitro,* of ingesting liposomes containing the missing enzymes. By this means we have been able to reconstitute the enzymic deficiencies of cells originally deficient in such enzymes as peroxidase and hexosaminidase A (the enzyme missing in Tay–Sachs disease) (Sessa and Weissmann, 1970; Weissmann *et al.*, 1975). Other investigators (Gregoriadis and Neerunjun, 1974) have studied the consequences of injecting enzyme-laden liposomes into living animals. A considerable number of the vesicles reach the lysosomal apparatus of appropriate cells and display enzymic activity sufficient to rid the intracellular digestive tract of some of its stored materials.

Unfortunately, even this approach is still in the earliest phases. We do not yet know, for example, how to direct these enzyme-containing vectors to cells that most need them, such as cells of the central nervous system, the heart, and the abdominal viscera. As yet, only early attempts have been made to breach the blood–brain barrier which seems to protect our central nervous system from picking up foreign materials from the bloodstream.

One can also insert materials into the surfaces of liposomes which will cause them directly to fuse with the plasma membrane of cells deficient in *cytoplasmic* (the cell sap) enzymes. One such agent, which tends to promote the fusion of biological membranes, is the fat lysolecithin. By means of varying the fatty nature of liposomal surfaces it has been possible to induce cells to accept their trapped enzymes after direct fusion of lysolecithin liposomes with the plasma membrane (Weismann *et al.*, 1977). Such experiments, as well as others from the laboratories of D. Papahadjopoulos (Poste and Papahadjopoulous, 1976) of Buffalo, have clearly demonstrated the introjection of the contents of liposomes into the free cytoplasmic juice of the cell.

Now this detailed account of new strategies for replacement of genetic enzyme deficiencies stands in marked contrast to the more radical proposals for direct gene therapy. Cytoplasmic palliation will not, of course, cure the basic defect inherent in the disordered genes themselves. It should therefore not be imagined that the approaches I have outlined are inherently preferable to, or should hold priority over, intense studies of more unique means to deliver DNA. As Lionel Trilling has suggested in another context, "I am inclined to suppose that whenever the genetic method is attacked we ought to suspect that special interests are being defended."

The special interest being defended here is the principle that all of our reparative eggs should perhaps not be put into the DNA recombinancy basket. There is no question but that most of the horrors imagined for DNA recombinancy research with mammalian cells are not only unlikely but very farfetched indeed. As Stanley Cohen (1977) has recently pointed out,

> Perhaps the single point that has been most misunderstood in the controversy about recombinant DNA research is that discussion of "risk" . . . relates entirely to hypothetical and speculative possibilities, not expected consequences or even phenomena that seem likely to occur on the basis of what is known. Unfortunately, much of the speculation has been interpreted as fact.

I would heartily agree, and would in fact urge that primary cure of genetic diseases might well eventually involve a combination of several approaches. However, before one discusses DNA recombinant research as a mode of entry into the therapeutic era of human genetic disease, we should perhaps pause to consider alternative approaches.

It may be argued that the palliative infusion of purified enzymes into patients missing them provides little benefit in the absence of radical intervention at the level of "the flaw in the gene." Granted, but the true cure of pneumococcal pneumonia would by these criteria require genetic eradication of pathogenicity in the pneumococcus. The true cure of tuberculosis would require elimination of those genetic sequences within the tubercle bacillus that code for elaboration of the toxin. However, we have been able to palliate pneumococcal pneumonia by the administration of penicillin and have been able to eliminate tuberculosis both by streptomycin and appropriate hygiene. Consequently, effective therapy in human disease does not necessarily depend on our intervention in the embryogenesis of disease, but may be brought about by prompt action in its adolescence.

So the argument returns again to feasibility, error, and risk. With respect to feasibility, I would suggest that the engineering techniques of enzyme-replacement therapy in mammals are much further developed than are those of the delivery of reparative genes. Problems of the uptake, degradation, targeting, and survival of infused enzymes, whether administered in free form or wrapped within layers, have been already approached and the parameters for judging success have been established. As to errors: these can be readily appreciated in experimental animals. Indeed, infusions of enzyme-containing liposomes have already been given to several humans with as yet no untoward reactions (albeit with no significant successes either!). As to risk, the risks involved in replacement therapy are unlikely to be transmitted, unlikely to be untreatable, and tend to recede as more work is done in experimental animals.

None of the articulate spokesmen for DNA splicing has yet publicly proposed that their technique is at all appropriate for the restitution of human (or animal) deficiency diseases, and so perhaps this whole argument is premature. But I doubt it. Much of the history of modern medicine is the application at the bedside of knowledge gained at the flask of *E. coli.* It may well turn out, with respect to the delivery systems involved, that all our hesitations and fears are groundless. But I am afraid there will have to be some new conceptual breakthrough, some new kind of feat of biological legerdemain before the spleens of patients with Gaucher's disease will shrink as the newly spliced genes start pumping enzymic iron. Genomic intervention in human disease is not impossible, nor is it, as far as we know, scientifically undesirable. At our present state of knowledge, however, there may perhaps be other alternatives. The study of these might well engage the attention of as many people as are now trying to persuade *E. coli* to manufacture insulin.

REFERENCES

Bangham, A. D., Standish, M. M., and Weissmann, G., 1965, The action of steroids and streptolysin S on the permeability of phospholipid structures to cations, *J. Mol. Biol.* **13**:253.

Brady, R. O., Tallman, J. F., Johnson, W. G., Gal, A. E., Lehy, W. R., Quirk, J. M., and Dekaban, A. S., 1973, Replacement therapy for inherited enzyme deficiency: Use of purified ceramidetrihexosidase in Fabry's disease, *N. Engl. J. Med.* **289**:9.

Brady, R. O., Pentchev, P., Gal, A., Hibbert, S., and Dekaban, A., 1974, Replacement therapy for inherited enzyme deficiency: Use of purified glucocerebrosidase in Gaucher's disease, *N. Engl. J. Med.* **291**:989.

Cohen, S. N., 1977, Recombinant DNA: Fact and fiction, *Science* **195**:654.

Gregoriadis, G., and Neerunjun, D. E., 1974, Control of the rate of hepatic uptake and catabolism of liposome-entrapped proteins injected into rats: Possible therapeutic applications, *Eur. J. Biochem.* **47**:179.

Gregoriadis, G., and Ryman, B. E., 1972, Fate of protein-containing liposomes injected into rats: An approach to the treatment of storage diseases, *Eur. J. Biochem.* **24**:485.

Gregoriadis, G., Leathwood, P. D., and Ryman, B. E., 1971, Enzyme entrapment in liposomes, *FEBS Lett.* **14**:95.

Hirschhorn, R., and Weissmann, G., 1977, Genetic disorders of lysosomes, *Prog. Med. Genet.* **1**:49.

Poste, G. and Papahadjopoulos, D., 1976, Lipid vesicles as carriers introducing materials into cultured cells: Influence of vesicle lipid composition on mechanism(s) of vesicle incorporation into cells, *Proc. Natl. Acad. Sci. USA* **73**:1603.

Ptashne, M., 1976, The defense doesn't rest, *The Sciences* **16**:11.

Sessa, G., and Weissmann, G., 1968, Phospholipid spherules (liposomes) as a model for biological membranes, *J. Lipid Res.* **9**:310.

Sessa, G., and Weissmann, G., 1970, Incorporation of lysozyme into liposomes: A model for structure-linked latency. *J. Biol. Chem.* **245**:3295.

Weissmann, G., Bloomgarden, E., Kaplan, R., Cohen, C., Hoffstein, S., Collins, T., Gottlieb, A., and Nagle, D., 1975, A general method for the introduction of enzymes, by means of immunoglobulin-coated liposomes, into lysosomes of deficient cells, *Proc. Natl. Acad. Sci. USA* **72**:88.

Weissmann, G., Cohen, C., and Hoffstein, S., 1977, Introduction of enzymes, by means of liposomes, into non-phagocytic human cells in vitro, *Biochim. Biophys. Acta* **498**:375.

Identifying Environmental Chemicals Causing Mutations and Cancer

BRUCE N. AMES

DAMAGE TO DNA BY ENVIRONMENTAL MUTAGENS AS A CAUSE OF CANCER AND GENETIC BIRTH DEFECTS

Damage to DNA by environmental mutagens (both natural and man-made) is likely to be a major cause of cancer (Doll, 1977; Hiatt *et al.*, 1977; Tomatis *et al.*, 1978) and genetic birth defects, and may contribute to heart disease (Benditt, 1977), aging (Burnet, 1974), cataracts (Jose, 1979), and developmental birth defects as well. Currently almost one-fourth of the population will develop cancer, and 5%–10% of children are born with birth defects. Damage to the DNA of germ cells can result in genetic defects that may appear in future generations. Somatic mutation in the DNA of the other cells of the body could give rise to cancerous cells by changing the normal cellular mechanisms, coded for in the DNA, that control and

---

BRUCE N. AMES • Department of Biochemistry, University of California, Berkeley, California 94720.

prevent cell multiplication. Exposure to mutagens occurs from natural chemicals in our diet, from synthetic chemicals (such as industrial chemicals, pesticides, hair dyes, cosmetics, and drugs), and from complex mixtures (such as cigarette smoke and contaminants in air and water).

A variety of data supports the hypothesis that environmental factors are a major cause of cancer (Doll, 1977; Hiatt *et al.*, 1977; Tomatis *et al.*, 1978). Epidemiological studies show different rates of incidence for certain types of cancer in different parts of the world. For example, in Japan there is an extremely low rate of breast and colon cancer and a high rate of stomach cancer, whereas in the United States the reverse is true. When Japanese immigrate to the United States, within a generation or two they show the high colon and breast cancer rates and low stomach cancer rates characteristic of other Americans. Known environmental mutagens that can cause human cancer include cigarette smoke tar, ultraviolet light, X-rays, and asbestos,* and the list of human chemical carcinogens is steadily lengthening (Doll, 1977; Tomatis *et al.*, 1978).

A high percentage of carcinogens are also likely to be able to reach and mutate the germ cells (Wyrobek and Bruce, 1975, 1978; Wyrobek and Gledhill, 1979), as well as the somatic cells, and the costs of this to society may be more than is generally realized.†

This chapter is not intended as a thorough review of what is clearly an enormous literature. I cite some recent reviews that describe the many important contributions that I am unable to discuss here and that contributed to our present ideas.

*Asbestos has recently been shown to be a potential mutagen: asbestos needles appear to pierce animal cells and cause chromosomal abnormalities (Sincock, 1977).

*E.g., smoking fathers have more babies with congenital abnormalities (Mau and Netter, 1974) and may have more abnormal sperm (Viczian, 1969) than nonsmoking fathers.

IDENTIFYING MUTAGENS AND CARCINOGENS: LIMITATIONS OF EPIDEMIOLOGY

Identifying the mutagens and carcinogens that cause cancer in people is tremendously difficult owing to a 20- to 30-year lag period between initial exposure to a carcinogen and the appearance of most types of human cancer. This is dramatically illustrated in the case of cigarette smoking (Figure 1). Men started smoking cigarettes in large numbers about 1900, but the resulting increase in lung cancer did not appear until 20–25 years later. Similarly, women started smoking in appreciable numbers about the time of World War II, and now the lung cancer rate for women is climbing rapidly. This same 20-year lag has been shown to apply for most types of cancer caused by the atomic bomb (leukemia and lymphoma show up earlier) and for cancer in factory workers exposed to a variety of chemicals. Cigarette smoking has been much easier to identify as a cause of cancer than most environmental carcinogens because there is a clear control group of nonsmokers, and because smoking causes a characteristic type of cancer (of the lung) that is infrequent in the control group. This is not the case with most environmental chemicals, and thus it is extremely difficult to convincingly identify the causal agent by epidemiology unless the background cancer incidence is increased by at least 100% (i.e., a doubling of risk).

Thus human epidemiology, although its use continues to be essential, cannot be our primary tool in detecting individual carcinogens because of the difficulties in connecting cause and effect (Peto, 1979), the great expense involved, and the fact that people would already have been exposed for decades by the time a particular cause of cancer was identified.

Human genetic defects are not easy to attribute to a specific cause, and a considerable increase in birth defects might go unnoticed. Moreover, many consequences of a gen-

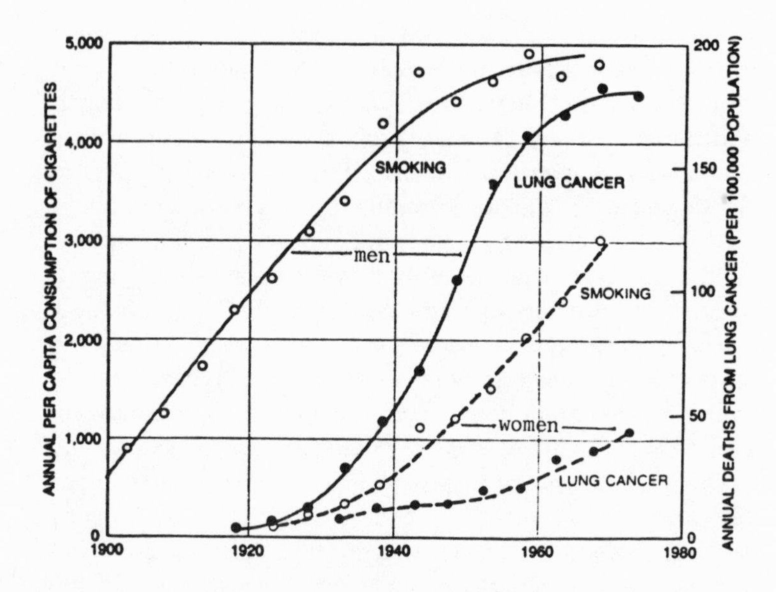

FIGURE 1. Relationship between cigarette smoking and lung cancer. Cigarette smoking and lung cancer are unmistakably related, but the nature of the relationship remained obscure because of the long latent period between the increase in cigarette consumption and the increase in the incidence of lung cancer. The data are for England and Wales. In men (solid line) smoking began to increase at the beginning of the 20th century, but the corresponding trend in deaths from lung cancer did not begin until after 1920. In women (dotted line) smoking began later, and lung cancers are only now appearing. (From Cairns, 1975, p. 72. Copyright 1975 by Scientific American, Inc. All rights reserved. Reprinted by permission.)

eral increase in gene mutations in the germ line might be subtle, such as decreased intelligence.

HUMAN EXPOSURE TO NEW CHEMICALS IN THE ENVIRONMENT

Clearly, many more chemicals will be identified as human mutagens and carcinogens. Currently over 50,000 synthetic chemicals are produced and used in significant quantities and close to 1000 new chemicals are introduced each year (Fishbein, 1977; Maugh, 1978). Only a small fraction of these were tested for carcinogenicity or mutagenicity *before* their use. In the past this problem was largely ignored, and even very high-production chemicals with extensive human exposure were produced for decades before adequate carcinogenicity or mutagenicity tests were performed. Such chemicals now known to be both carcinogenic and mutagenic include vinyl chloride (produced at a rate of about 6 billion lb per year in the U.S. in 1977) and 1,2-dichloroethane (ethylene dichloride, about 10 billion lb per year; Figure 2), and a host of high-production pesticides.

The increase in production and use of chemicals has been particularly great since the mid-1950s, as illustrated by the two high-production compounds shown in Figure 2. This flowering of the chemical age may be followed by genetic birth defects and a significant increase in human cancer during the 1980 decade (because of the 20- to 30-year lag) if many of these chemicals with widespread human exposure are indeed powerful mutagens and carcinogens.

Some of these environmental carcinogens accumulate in human body fat; almost all of us are continuously being exposed to low, but disturbing, dose levels of these carcinogens. Table 1 shows some chlorine-containing chemicals found in

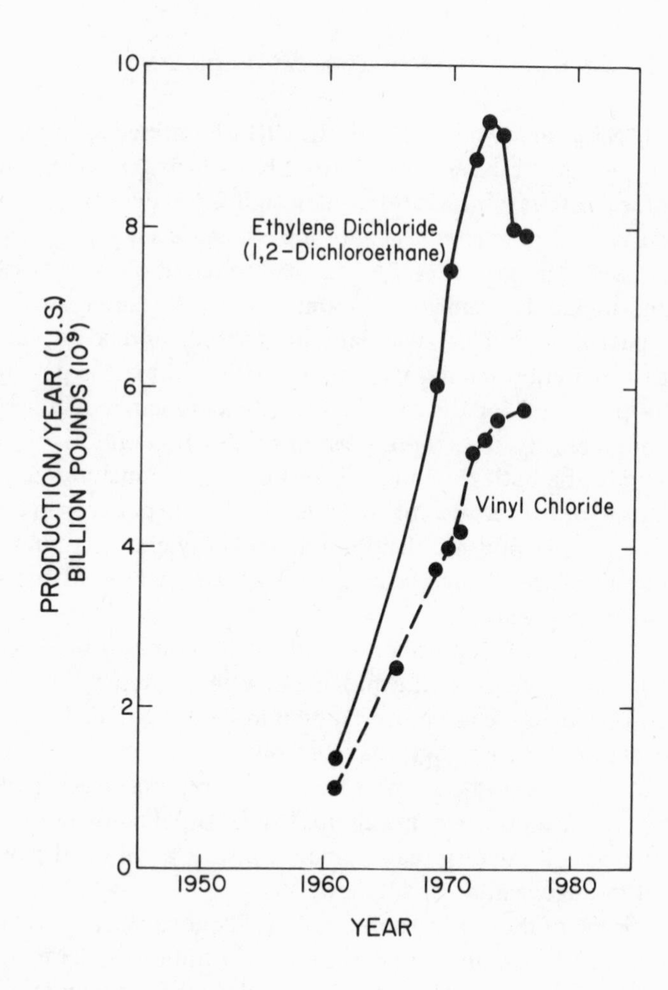
10
8
6
4
2
0
PRODUCTION/YEAR (U.S.)
BILLION POUNDS (10⁹)
Ethylene Dichloride
(1,2-Dichloroethane)
Vinyl Chloride
1950
1960
1970
1980
YEAR

FIGURE 2. Production of two mutagens–carcinogens with widespread human exposure: ethylene dichloride and vinyl chloride (production data from "Top-50 Chemicals" issues of *Chemical and Engineering News*). Approximately 100 billion lb (5×10^{10} kg) of ethylene dichloride and over 50 billion lb of vinyl chloride have been produced since 1960. Ethylene dichloride is a volatile liquid that is the precursor of vinyl chloride and is also used extensively as a fumigant, solvent, gasoline additive (200 million lb per year), and metal degreaser. Ethylene dichloride was first shown to be a mutagen in *Drosophila* in 1960 (Rapoport, 1961; Shakarnis, 1969), and later in barley and *Salmonella*, but these data have been ignored. The first adequate cancer test in animals has just been completed by the National Cancer Institute (September 1978) and is positive in both sexes of both rats and mice. Vinyl chloride gas is used to make polyvinyl chloride (PVC; vinyl) plastic. It was shown to be a carcinogen in rats and in people in the mid-1970s, and a mutagen in *Salmonella* and other systems shortly afterward. Vinyl chloride production results in dumping enormous quantities of a waste product, EDC-tar, that is a complex mixture of chlorinated hydrocarbons. This is quite mutagenic in *Salmonella* and has also been detected as an ocean pollutant (Rannug and Ramel, 1977).

TABLE 1. Chlorinated Hydrocarbon Residues in Human Fat (Average of 168 Canadian Samples)[a,b]

Compound	Mean wet weight (μg/kg)	% of samples containing residues
PCB	907	100
Hexachlorobenzene	62	100
BHC (Lindane)	65	88
Oxychlordane	55	97
Trans-nonachlor	65	99
Heptachlor epoxide	43	100
Dieldrin	69	100
p,p'-DDE	2095	100
o,p'-DDT	31	63
p,p'-TDE	6	26
p,p'-DDT	439	100

[a]Adapted, with permission, from Mes *et al.* (1977). Copyright by Springer-Verlag, New York.
[b]Almost all have been shown to be carcinogens.

TABLE 2. Some Pesticides in Human Milk[a]

Pesticide	Number positive (%)	Mean of positives (μg/kg fat)[c]	Maximum (μg/kg fat)
DDE	100	3,521	214,167
DDT	99	529	34,369
Dieldrin	81	164	12,300
Heptachlor epoxide	64	91	2,050
Oxychlordane	63	96	5,700
β-BHC	87	183	9,217
PCBs	30[b]	2,076	12,600

[a]1400 women, Environmental Protection Agency data, 1976; PCBs, 1038 women, Environmental Protection Agency data, 1977. Adapted with permission from Harris and Highland, 1977. Data assembled from E. P. Savage (1976).
[b]99% detectable PCBs (30% = > 1100 μg/kg fat).
[c]4.5% mean fat content.

human body fat: almost all are known carcinogens in rodents. Table 2 shows the levels of some of these same carcinogens in human mothers' milk.

The few chemicals that were assayed for in these studies are only some of those that have accumulated in people. Carcinogens such as toxaphene (Hooper *et al.*, 1979), kepone, mirex, and other major chlorinated pesticides are also accumulating in the food chain, as are a wide variety of industrial chemicals as yet untested for carcinogenicity. Some of these, like the notorious polybrominated biphenyls (PBBs), are very likely to be carcinogens.*

Organic chemicals containing chlorine and bromine are not used in natural mammalian biochemical processes and may not have been usually present in the human diet until the onset of the modern chemical age. A very high percentage of chlorinated and brominated chemicals are carcinogens in animal cancer tests and thus represent a highly suspect class of chemicals.

IS THERE A SAFE DOSE OF MUTAGENS AND CARCINOGENS?

There is considerable debate as to whether there is a safe dose for carcinogens and whether we should be concerned about the exposure of the general population to the many low doses of environmental carcinogens. There is no firm answer

*About 500 lb of the polybrominated biphenyls (PBBs) were accidentally mixed into Michigan cattle feed in 1973, causing contamination of dairy farms in the state. Millions of pounds of this chemical have been produced, and its environmental dispersion may eventually haunt us as the polychlorinated biphenyls have done (Carter, 1976). Pentachlorophenol, the most commonly used wood preservative, is found in human semen at low levels (Dougherty and Piotrowska, 1976). Among the many compounds identified in the expired air of normal adults are five different chlorinated compounds, including dichlorobenzene (used as moth crystals; Krotoszynski *et al.*, 1977).

to this question since it is impossible to get statistically significant information at very low dose levels from animal cancer experiments in which only a few hundred animals are exposed to the chemical. Several arguments suggest that thresholds, or completely safe doses of carcinogens, are not likely to be the general case,* and it seems prudent to assume, until shown otherwise experimentally, that small increments of an environmental carcinogen will increase risk linearly.

THE UTILITY AND LIMITATIONS OF ANIMAL CANCER TESTS

A key method for detecting carcinogens is the animal bioassay, usually with rats and mice. Almost all of the dozen or so organic chemicals known to cause cancer in humans also cause cancer in experimental animals when adequately tested (Doll, 1977; Tomatis *et al.*, 1978). A fairly small percentage of chemicals in general appear to be carcinogens. The National Cancer Institute has just completed testing about 200 suspicious industrial compounds to which people are exposed in appreciable amounts. Many more such tests need to be done. The utility of animal cancer tests for cancer prevention, however, is limited by several important factors.

Feasibility: Time and Expense. Animal cancer tests are too expensive (currently about $250,000 per chemical for a

*Dose–response information from humans on lung cancer incidence and cigarette smoking (Doll, 1978) does not give a clear-cut answer, although the best fit is quadratic with a low (1.7) exponent for dose. Studies on humans exposed to radiation also provide no support for the safe dose concept. In addition, because we are high up on the human dose–response curve (25% of us will get cancer), and because most carcinogens appear to have the same mechanism of action as mutagens, it has been argued that the doses of most carcinogens to which we are exposed could have an additive effect and increase risk linearly (Crump *et al.*, 1976; Guess *et al.*, 1977), although in particular cases synergisms or interferences may result.

thorough test) and take too long (about 3 years) to be used for the testing of the many thousands of chemicals to which humans are exposed. In fact, adequate animal cancer tests are reported for only about 150 previously untested chemicals each year. There are not enough pathologists to read the slides even if we decided to test only the thousand or so new chemicals introduced into commerce each year, not to mention the 50,000 untested commercial chemicals already in use and the even-greater number of chemicals in the natural world. Animal tests are also not practical as bioassays for identifying the carcinogens in the many complex chemical mixtures that surround us, such as natural chemicals in our diet, cigarette smoke, impurities in water and air, and complex industrial products. A further limitation is that chemical and drug companies need to have a method for weeding out hazardous chemicals while they are still under development and while alternatives can be substituted. Currently, many chemicals undergo a long-term animal test, if they are tested at all, only after millions of dollars have been invested in them.

Sensitivity. An environmental carcinogen causing cancer in 1% of 100 million people would result in a million cases of cancer. To detect a chemical causing cancer in only 1% of the test animals, we would have to use 10,000 rats or mice, which would be extraordinarily expensive, and in fact a test group of only 50 mice or rats of each sex at each of two doses is the usual size of the most thorough cancer experiments. This limitation is somewhat overcome, although not entirely satisfactorily, by exposing the animals to as high a dose as possible (the "maximum tolerated dose") which, by increasing the tumor incidence, partially offsets the statistical problems inherent in the small sample size.

THE *SALMONELLA*–MAMMALIAN LIVER TEST: DETECTING CHEMICAL MUTAGENS

Over the past 15 years we have been developing a simple test for identifying chemical mutagens* and have used it to show that about 90% of organic chemical carcinogens tested thus far are mutagens (Donahue *et al.*, 1978; McCann and Ames, 1976, 1977; McCann *et al.*, 1975). This work, and that of others (Hiatt *et al.*, 1977; Maher *et al.*, 1968; Miller and Miller, 1977; Nagao *et al.*, 1978; Purchase *et al.*, 1976, 1978; Sugimura *et al.*, 1976), have strongly supported the theory that most carcinogens act by damaging DNA. The *Salmonella* test and other short-term tests that have been developed based on testing chemicals for ability to interact with DNA or for mutagenicity† are being widely used and should help in solving some of the problems that cannot adequately be approached by human epidemiology or animal cancer tests alone. The test is in current use in over 2000 government, industrial, and academic laboratories throughout the world. A number of companies have made important economic decisions on the basis of this test.**

Our work started as an offshoot from our basic research on the molecular biology of *Salmonella* bacteria. We were studying how genes are switched on and off in bacteria in response to the presence of histidine in the growth medium and the effect of mutations which perturbed this control mechanism. We were using a large collection of histidine-requiring bacterial mutants, mostly made by Philip E.

*See Ames *et al.* (1975b) for a description of the method and references to earlier work.

†Hollstein *et al.* (1979) have recently reviewed the many short-term tests.

**The Dupont chemical company decided (at considerable economic loss) not to produce two freon propellants for use in spray cans because they were found to be mutagenic (these were available replacements for the freon that is thought to damage the ozone layer).

Hartman of Johns Hopkins University. In 1964 we began to develop a test system for detecting mutagens using these mutants. During the course of this work on test development and on the theory of frameshift mutagenesis, we became convinced that the essential property of most carcinogens was their mutagenicity (Ames, 1971; Ames and Whitfield, 1966; Ames *et al.*, 1972a,b). The ideas and work of the Millers, Boyland, Magee, the Weisburgers, and other workers (Maher *et al.*, 1968; Miller and Miller, 1977), had made a major contribution to the understanding that many carcinogens must be converted by enzymes in liver or other tissues to an active (electrophilic) form that is the true carcinogen (and mutagen). Also of great importance was the *in vitro* work from the laboratories of Malling, Rosenkranz, and the Millers on the use of liver homogenate for the activation of dimethylnitrosamine, diethylnitrosamine and aflatoxin combined with our tester strains, or *E. coli*, as indicators of DNA damage (Garner *et al.*, 1972; Malling, 1971; Slater *et al.*, 1971). We thus added mammalian liver tissue to the test to provide a first approximation of mammalian metabolism.

The test is done by combining on a petri plate the compound to be tested, about 1 billion *Salmonella* bacteria of a particular tester strain (several different histidine-requiring mutants are used), and homogenized liver from rodents (or human autopsy); after incubation at 37°C for 2 days, the number of bacterial revertant colonies is recorded. Each colony is composed of the descendants of a bacterium that has been mutated from a defective histidine gene to a functional one. Normally, one tests doses of a chemical in a series of plates—the plate test—and quantitative dose–response curves are generated (Figure 3). These curves are almost always linear, which suggests that, at least in this system, thresholds are not common. Mutagens can be detected at low doses, in some cases in nanogram amounts.

Validation: Testing 300 Chemicals. We validated the test for

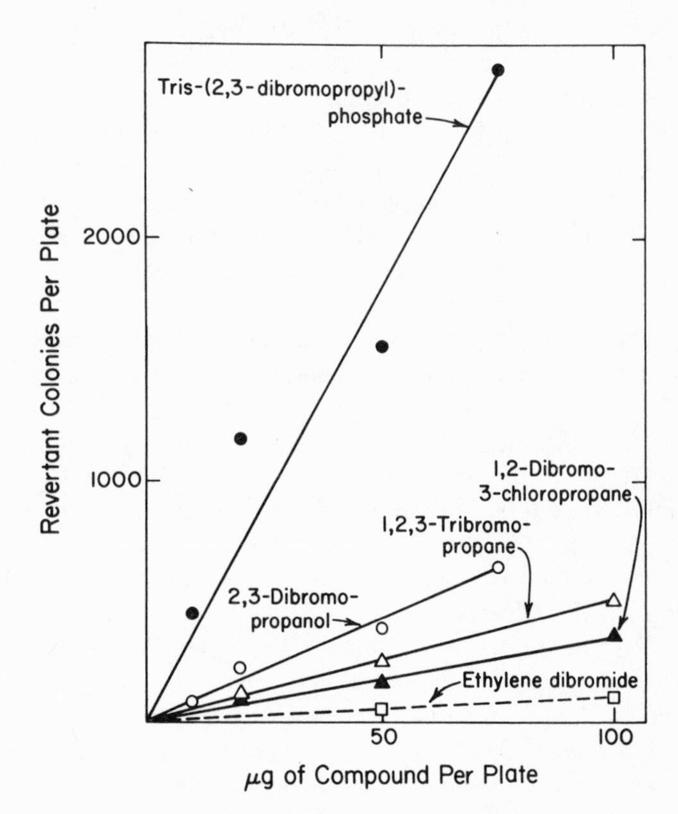

FIGURE 3. The plate test quantitative assay: linear dose responses. The flame retardant *tris*-(2,3-dibromopropyl) phosphate, its metabolite in children dibromopropanol, and the pesticide dibromochloropropane were in the presence of rat liver homogenate. All compounds were tested on *Salmonella* strain TA100. The amount of the industrial chemical ethylene dibromide added was 10 times that indicated on the scale. (From Blum and Ames, 1977; reprinted with permission.)

the detection of carcinogens as mutagens by examining about 300 chemicals reported as carcinogens or noncarcinogens in animal experiments (Donahue *et al.*, 1978; McCann and Ames, 1976, 1977; McCann *et al.*, 1975). About 90% (158 of 176) of these chemical carcinogens were mutagenic in the *Salmonella* test. Our test system has subsequently been independently validated, with similar results (Purchase *et al.*, 1976, 1978; Sugimura *et al.*, 1976). The percentage of carcinogens detectable would, of course, depend on how nearly any particular list of carcinogens was representative of those existing in the real world. For this reason we also examined the organic chemicals known or suspected as human carcinogens and found that almost all (16 of 18; benzene and diethylstilbestrol, or DES, were the exceptions) were mutagens in the test (McCann and Ames, 1976, 1977; McCann *et al.*, 1975). Nevertheless, it is important to reemphasize that the success rate is markedly lower* for some classes of carcinogens such as hydrazine and heavily chlorinated chemicals (McCann and Ames, 1976, 1977; McCann *et al.*, 1975). Even with further test improvements, promoters and some carcinogens (e.g., griseofulvin and steroids) may never be detected because they probably do not act through a direct interaction with DNA (McCann and Ames, 1976, 1977; McCann *et al.*, 1975).

Thus, a very high percentage of carcinogens tested are mutagens, and most mutagens appear to be carcinogens. We found that few (13 of 108) "noncarcinogens" tested were mutagenic, and these few may in fact be weak carcinogens that were not detected as such due to the statistical limitations of animal carcinogenicity tests (Donahue *et al.*, 1978; McCann and Ames, 1976, 1977; McCann *et al.*, 1975). Only a small

*We have pointed out (McCann and Ames, 1976, 1977; McCann *et al.*, 1975) that it is important to view with caution results indicating nonmutagenicity in classes of chemicals for which the test (or any short-term test) has not yet been validated.

percentage of chemicals in general appears to be mutagens (McMahon *et al.*, 1979).

Mutagens Subsequently Found to Be Carcinogens. There have been a number of instances of chemicals initially detected as mutagens that have subsequently been found to be carcinogens (Miller *et al.*, 1979; Nagao *et al.*, 1978). I will discuss five of these that involved extensive human exposure.

1. AF-2 (furylfuramide) was a food additive used extensively in Japan from 1965 until recently as an antibacterial additive in a wide variety of common food products such as soybean curd and fish sausage (Inui *et al.*, 1978; Nagao *et al.*, 1978). It showed no carcinogenic activity in tests on rats in 1962 and on mice in 1971. In 1972 and 1973, however, Japanese scientists found that it caused chromosomal aberrations in cultured human cells and was also highly mutagenic in a strain of *Escherichia coli* bacteria. It was later found to be an extraordinarily potent mutagen in our *Salmonella* test, so much so that one could easily demonstrate the mutagenicity of a slice of fish sausage put on a petri plate (Nagao *et al.*, 1978). It was later found to be mutagenic in yeast and *Neurospora* and in a variety of other short-term tests, and more recently to mutate embryos when even low doses were fed to pregnant Syrian hamsters (Inui *et al.*, 1978). More thorough animal tests for carcinogenicity were initiated, and these tests have recently shown that AF-2 is in fact a carcinogen in rats, mice, and hamsters. As a consequence, the Japanese government prohibited the use of AF-2 as a food additive, and all products containing AF-2 were removed from the market (Nagao *et al.*, 1978).

Since AF-2 had already been tested for carcinogenicity in two animal systems and found negative, it is unlikely that further animal tests would have been conducted if it had not been shown to be mutagenic; and any deleterious effects on the Japanese population would probably not have been evident for decades. The Japanese people did consume relatively large amounts of furylfuramide for 9 years, and it is still too early to assess the consequences of this exposure.

2. Ethylene dichloride, a 10-billion-lb-per-year chemical (see Figure 2) first shown to be a mutagen in the fruit fly *Drosophila,* and then in barley and in *Salmonella,* was recently tested by NCI and found to be a carcinogen in both sexes of rats and mice.

3. Ethylene dibromide (1,2-dibromoethane; Fishbein, 1977; Maugh, 1978) is widely used (400 million lb per year in the U.S.) industrial chemical and gasoline additive, which was detected as a mutagen in *Neurospora* in 1969 (Malling, 1969) and then in several microbial systems (including *Salmonella;* see Figure 3). In 1973 it was tested for carcinogenicity and found to be quite a potent carcinogen in rats and mice.

4. The flame-retardant "tris-BP" [*tris*-(2,3-dibromopropyl)phosphate], a related dibromo chemical which was the main flame retardant in children's polyester pajamas, is a mutagen in our test system; so are its metabolic product dibromopropanol, and its impurity, the carcinogen (and human sterilant) dibromochloropropane (DBCP; see Figure 3; Blum and Ames, 1977; Prival *et al.,* 1977). During the years 1972–77, 50 million children wore sleepwear that contained this material, at about 5% of the weight of the fabric. We argued that tris-BP would pose a serious hazard to children because nonpolar (relatively fat-soluble and water-insoluble) chemicals such as tris-BP are generally absorbed through human skin at appreciable rates (Blum and Ames, 1977; Prival *et al.,* 1977). Since its detection as a mutagen in *Salmonella,* it has been shown to be active in a number of short-term tests: it is a potent mutagen in *Drosophila,* it interacts with human DNA, and it damages mammalian chromosomes. The compound was tested recently at the National Cancer Institute and was shown to be a carcinogen in both rats and mice. It has also

been shown to cause cancer in skin-painting studies on mice (van Duuren *et al.*, 1978), and, like DBCP, to cause sterility in animals. It is no longer being used in sleepwear. We have recently shown that a mutagenic metabolite of tris-BP, dibromopropanol, is present in the urine of children wearing tris-treated sleepwear (Blum *et al.*, 1978).

5. Hair dyes have also been shown to contain mutagens. In a study at our laboratory (Ames *et al.*, 1975; more recent work is reviewed by Marzulli *et al.*, 1978), about 90% (150 of 169) of commercial oxidative-type (hydrogen peroxide) hair dye formulations were found to be mutagenic, and of the 18 components of these hair dyes (mostly aromatic amines), 8 were mutagenic. Many semipermanent hair dyes tested were also shown to be mutagenic. Hair dye components are known to be absorbed through the skin, yet very few of the hair dyes, their components, or their peroxide reaction products have ever been tested adequately for carcinogenicity. Since the work on mutagenicity in *Salmonella,* a variety of these ingredients have been shown to be mutagens in other short-term tests. Several of the chemicals are being tested at the National Cancer Institute and now appear to be carcinogens. About 25 million people in the United States (mostly women) dye their hair, and the hazard could be considerable if these chemicals are mutagenic and carcinogenic in humans. A recent epidemiological study suggests that there may be a considerable excess of breast cancer in postmenopausal women who have dyed their hair over a long period (Shore *et al.*, 1979), although more definitive work needs to be done.

Identifying Mutagenic Components in Complex Mixtures. The sensitivity of the *Salmonella* test makes it useful for rapidly obtaining information about mutagenic components of complex mixtures such as impurities in industrial chemicals (Donahue *et al.*, 1978), natural products, air and water pollutants, pyrolyzed material, and body fluids (Nagao *et al.*, 1978).

For example, a detailed study has been made of the mutagenic activity of cigarette smoke condensate and 12 standard smoke condensate fractions (Kier *et al.*, 1974; Nagao *et al.*, 1978). In the test, the condensate from less than 0.01 cigarette could easily be detected.

We have recently developed a simple method for examining human urine in our test system and have found mutagens in the urine of cigarette smokers but not (at the level of sensitivity used) in the urine of nonsmokers (Yamasaki and Ames, 1977).

NATURAL CARCINOGENS AND MUTAGENS IN THE DIET

Any major effect of man-made chemicals as carcinogens and mutagens should become apparent in the next several decades. Much of the cancer we see today, on the other hand, in addition to that caused by cigarette smoke and radiation (such as ultraviolet light which induces skin cancer), appears likely to be due to the ingestion of a wide variety of natural carcinogens in our diet (Miller *et al.*, 1979). Fat intake has been correlated with breast and colon cancer (Hiatt *et al.*, 1977; Miller *et al.*, 1979). Plants have developed a wide assortment of toxic chemicals (probably to discourage insects and animals from eating them) and many of these are mutagens and carcinogens that are present in the human diet (Bjeldanes and Chang, 1977; Brown *et al.*, 1977; Hashimoto *et al.*, 1979; Hiatt *et al.*, 1977; Miller *et al.*, 1979; Nagao *et al.*, 1978). In addition, powerful nitrosamine and nitrosamide carcinogens are formed from certain normal dietary biochemicals containing nitrogen, by reaction with nitrite (Bruce *et al.*, 1977; Lijinsky and Taylor, 1977; Miller *et al.*, 1979; Wang *et al.*, 1978). Nitrite is produced by bacteria in the body from nitrates that are present in ingested plant material and water (Hiatt *et al.*, 1977; Lijinsky and Taylor, 1977; Miller *et al.*, 1979; Wang *et al.*, 1978). A number of molds produce powerful carcinogens such as aflatoxin and sterigmatocystin that can be present in small amounts in food contaminated by molds, such as peanut butter and corn (Hiatt *et al.*, 1977; Miller *et al.*, 1979).

Several studies of major importance are in progress using the *Salmonella* test to identify natural carcinogens. Dr. W. R. Bruce and his colleagues in Toronto have found a considerable amount of a powerful mutagen in human feces (Bruce *et al.*, 1977; Miller *et al.*, 1979). It appears to be a nitrosamide-type compound formed from a component of dietary fat and nitrite and could be a cause of colon and breast cancer, two common cancer types associated with high fat intake. Bruce is identifying its chemical structure using *Salmonella* mutagenicity as a bioassay. He has also found that ingesting vitamin C or vitamin E lowers the amount of the mutagen present in feces (Miller *et al.*, 1979).

In another instance, using the *Salmonella* test as a bioassay, Sugimura and other workers in Japan have discovered that when fish or other foods containing protein are broiled, mutagenic chemicals are formed (Miller *et al.*, 1979; Nagao *et al.*, 1977, 1978; Yoshida *et al.*, 1978). They have also found that broiling protein produces mutagens and that broiling tryptophan (a component of protein) produces extremely potent mutagens. Several mutagens have been identified chemically using *Salmonella* as a rapid bioassay, and one has been shown to be extremely active in another short-term assay (transformation) using animal cells (Miller *et al.*, 1979). Animal cancer tests are being done on the pure substances. (Identifying the active chemicals by the use of animal cancer tests would have been impractical because of the time involved.) Commoner *et al.* (1978) have reported that fried hamburgers also show mutagenic activity in the *Salmonella* test.

Glycosides of quercetin, a mutagenic flavonoid, are present in considerable amounts in our diet from a variety of sources (Bjeldanes and Chang, 1977; Brown *et al.*, 1977; Miller *et al.*, 1979; Nagao *et al.*, 1978), and bacteria in the human gut readily hydrolyze off the sugars to liberate the mutagen. The contribution to human cancer of mutagenic flavonoids and anthraquinone and other plant mutagens present in our diet remains to be evaluated (Bjeldanes and Chang, 1977; Brown *et al.*, 1977; Hashimoto *et al.*, 1979; Miller *et al.*, 1979; Nagao *et al.*, 1978).

OTHER SHORT-TERM TESTS FOR MEASURING MUTAGENS

With the development of the *Salmonella* test and the demonstration that in general carcinogens are mutagens, there has been a tremendous surge of interest in the other short-term test systems for measuring mutagenicity. Many such systems have been developed (Hiatt *et al.*, 1977; Hollstein *et al.*, 1979; Nagao *et al.*, 1978) and have made major contributions to the field. Some, including the use of animal cells in tissue culture which can be examined both for mutagenic and cytogenetic damage, have been validated by testing a number of carcinogens and noncarcinogens. In addition, a number of the older systems, such as mutagenicity testing in *Drosophila,* have been improved. (The first mutagens known, such as X-rays and mustard gas, were identified in *Drosophila* before they were known to be carcinogens.) An important advance is the development of several tissue-culture systems with animal cells, having as an end point the "transformation" of the cells to cells that can form tumors (Hollstein *et al.*, 1979) when injected into animals. A number of tests with rodents also are being developed to examine mutagenic damage in cells in the whole animal (Hiatt *et al.*, 1977; Hollstein *et al.*, 1979; Inui *et al.*, 1978).

No single short-term test, however, is perfect. Because each system detects a few carcinogens that others do not, the idea of a battery of short-term tests is now favored by many investigators (Hollstein *et al.*, 1979). It is becoming apparent that positive results from a battery of these short-term test systems are meaningful; these systems, as well as complementing animal cancer tests, provide useful toxicological information about mutagenicity.

Mutagens should be treated with respect not only because of their probable carcinogenicity, but also because of other biological consequences of DNA damage, such as genetic birth defects. There are now simple methods for looking at effects of chemicals on the germ line (Hollstein *et al.*, 1979; Wyrobek and Bruce, 1975, 1978; Wyrobek and Gledhill, 1979), such as testing for sterility or defective sperm in rodents or people (Wyrobek and Bruce, 1975, 1978; Wyrobek and Gledhill, 1979). Because of the discovery that the carcinogen and mutagen dibromochloropropane (DBCP) causes sterility in both rodents and male workers* and causes chromosomal abnormalities in the sperm of chemical plant workers, and because of the finding that a high percentage of carcinogens damage the germ line as well (Mau and Netter, 1974; Viczian, 1969; Wyrobek and Bruce, 1975, 1978; Wyrobek and Gledhill, 1979), interest in these methods should increase.

CARCINOGENIC POTENCY AND HUMAN RISK ASSESSMENT

With the large number of environmental carcinogens and mutagens, both man-made and natural, it is clearly impractical

*The pesticide dibromochloropropane (DBCP) was used until recently at a level of about 10 million lb per year in the U.S. In 1961 it was shown to cause sterility and testicular atrophy in animals (Torkelson *et al.*, 1961), in 1973 it was shown to be a carcinogen (Olson *et al.*, 1973), and in 1977 it was shown to be a mutagen in *Salmonella* (Blum and Ames, 1977; Prival *et al.*, 1977). Its potency as a carcinogen (Sawyer, C., Hooper, N. K., and Ames, B. N., unpublished calculations done on NCI cancer test) is such that the daily dose (mg/kg) to give 50% of the male and female rats cancer (it is only slightly less potent in male and female mice) is approximately the daily exposure level of a worker breathing air contaminated with 2 ppm of DBCP—close to the actual level of worker exposure. It has been shown recently to cause abnormalities in the sperm of workmen (Kapt *et al.*, 1979).

to ban or eliminate every one. We must have some way of setting priorities for dealing with these chemicals—and this requires an assessment of human risk, a difficult and complex problem.

A knowledge of carcinogenic potency would be an important aid in human risk assessment. Following the lead of M. Meselson (Meselson and Russell, 1977) and collaborating with R. Peto on the theoretical aspects, we (N. K. Hooper, C. B. Sawyer, A. D. Friedman, R. H. Harris, and B. N. Ames) have been working on the potency problem for several years and are nearing completion of our quantitative analysis of the several thousand published animal cancer tests in which a chemical was fed continuously for an appreciable fraction of the lifetime of the animal. We have shown that the potency of carcinogens (the TD_{50}, the daily dose required to produce cancer in half of the animals, or more precisely, to reduce the probability of being tumor-free by one-half) can vary by well over a millionfold. Such a range of potency must be considered in assessing the hazard of chemicals for man and shows that it is essential to consider carcinogens in more quantitative terms. These results should be useful in determining the following:

1. Which chemicals, among the thousands of carcinogens to which people are exposed, are likely to present the greatest human hazards and thus require the most immediate attention. Setting priorities also requires, of course, an estimate of the amount of human exposure to a given chemical.
2. Better ways of calculating unacceptable levels of carcinogen exposure for workers or the general population.
3. The significance of negative animal cancer tests. Each particular cancer test has a particular thoroughness (sensitivity)—because of the dose level of chemical used, numbers of animals, and other factors—and can

detect only those carcinogens having potencies above a certain level. Because cancer tests vary enormously in thoroughness, we have expressed the results of a negative cancer test by assigning the chemical a maximum potency value rather than using the quantitatively meaningless term "noncarcinogen."

4. The extent to which carcinogenic potency is species, strain, and sex specific, and similar in long-lived species such as monkeys and short-lived rodents. Our analysis so far indicates that, in general, potency values for a given chemical do not vary much between males and females, or, with a few significant exceptions, between rats and mice.

We believe that the potency scale can be applied to a number of current problems, such as human exposure limits for volatile carcinogens in the workplace (threshold limit values). These TLVs might be safer and more rationally determined if carcinogenic potency is taken into account. Under the present standard-setting system, workers are allowed to breathe in amounts of certain volatile halogen-containing solvents, e.g., DBCP or ethylene dichloride (Sawyer *et al.*, unpublished calculations). The establishment of a priority list for investigating these and other chemicals that have wide use and an appreciable carcinogenic potency is an urgent necessity, as is the examination of workers and other exposed populations for the effects of these chemicals.

POTENCY IN SHORT-TERM MUTAGENICITY TESTS: CORRELATING *SALMONELLA* AND ANIMAL TEST DATA

Although a start can now be made on human risk assessment based on animal cancer tests, few of the chemicals to which people are exposed in the environment have actually undergone cancer testing in animals. Furthermore, many of the completed tests lack the quality needed to make a quantitative analysis of the data. Thus the question remains: Can short-term tests provide any quantitative information about human risk?

There is more than a millionfold range in mutagenic potency in the *Salmonella* test and there is also a similar range in carcinogenic potency (McCann and Ames, 1976, 1977; McCann *et al.*, 1975; Nagao *et al.*, 1978). Although one would certainly not expect a precise quantitative correlation between mutagenicity in a bacterial liver test and carcinogenicity in animals, even a rough quantitative correlation would be useful in human risk assessment. Work done by Meselson and Russell (1977) on 14 chemicals suggests that there is a good correlation of the two potencies, not only for carcinogens in the same class, but also across a broad range of classes, although some nitrosamines did not fit this general relationship.

We are comparing the potency of chemicals in causing tumors in rats with potency in the *Salmonella* test (using a rat liver homogenate for activation) (N. K. Hooper, A. D. Friedman, C. B. Sawyer, and B. N. Ames, manuscript in preparation; Ames and Hooper, 1978). The results so far are promising and additional work will show how general this correlation is.* It is feasible to obtain *Salmonella* mutagenicity data on all those carcinogens for which one can calculate a carcinogenic potency. Our analysis should, in any case, give some indication of the chemicals for which it is necessary in mutagenicity tests to use tissue homogenates other than from liver and of areas where the test needs improvement.

*The theoretical basis for a correlation may be that using rat liver homogenate as a model for a rat's metabolism of foreign chemicals is a reasonable first approximation for ingested chemicals. The liver is the primary organ for metabolizing foreign chemicals and in general is much more active than other tissues for metabolic activation. While most carcinogens that are ingested do not cause liver cancer, this may be explainable by the increased DNA-repair capabilities of the liver (Denda *et al.*, 1977; Kleihues and Bucheler, 1977).

A number of laboratories are examining to what extent species differences in carcinogenic potency of chemicals can be correlated with differences in mutagenic potency by using the liver or other tissue homogenates from the different species. Other short-term tests that are currently being developed can also be calibrated against our carcinogenic potency index to see how well they correlate. The quantitative agreement between *Salmonella* and another short-term test—inhibition of DNA synthesis in human (HeLa) cells in tissue culture—has been recently examined and appears good (Painter and Howard, 1978). If several short-term tests can be shown to provide rough quantitative results consistent with those from animal cancer tests, a battery of short-term tests could then be used to help establish priorities among the many mutagens, both natural and man-made, that have never been tested in animal cancer tests and to which there is significant human exposure.

THE PREVENTION OF DNA DAMAGE

The problems of cancer, genetic birth defects, and other consequences of DNA damage can be usefully attacked by prevention. The following approaches (I will not discuss regulatory policy) are suggested:

1. *Identifying mutagens and carcinogens* from among the wide variety of environmental chemicals to which humans are exposed. All approaches must be used: human epidemiology for cancer and genetic birth defects; animal tests for cancer and for genetic birth defects; and short-term mutagenicity and transformation tests.
2. *Premarket testing* of new chemicals to which humans will be exposed.
3. *Making information more readily available* on both natural

and man-made chemicals capable of causing cancer and mutations (including their relative danger where this is known) for use by the state and federal governments, industry, unions, consumer groups, and the public at large.*

4. *Establishing priorities for trying to minimize human exposure* to these chemicals. Priority lists could take into account, among other factors, amount of human exposure to each chemical and the potency of the chemical in animal cancer tests.† Where adequate animal cancer data will not soon be available, potency information from *several* suitable short-term tests might be substituted provided that the tests have been validated for this purpose. It seems unlikely that animal cancer tests can soon catch up with the many mutagenic substances being discovered (Nagao *et al.*, 1978). Soon, more sophisticated and sensitive ways of measuring DNA or other damage by mutagenic chemicals in people may play an essential role in risk assessment (Ehrenberg *et al.*, 1977; Segerbäck *et al.*, 1978; Wyrobek and Bruce, 1975, 1978; Wyrobek and Gledhill, 1979).
5. *Discovering modifying factors* in carcinogenesis such as vitamins C and E (Bruce *et al.*, 1977; Miller *et al.*, 1979), genetic factors (e.g., skin color in ultraviolet carcinogenesis; Mulvihill *et al.*, 1977), selenium (Dillard *et*

*Some individuals affected, when told the risks, may judge many carcinogens (particularly ones they are used to) to be tolerable if the risk is low. The risk does not deter smokers, even those knowing that the average two-pack-a-day smoker has a life expectancy about 8 years less than the average nonsmoker (Gail, 1975).

†Mixtures such as air pollutants from automobile exhausts, including diesel (both are mutagenic in *Salmonella*), may be on the priority list. A general attack on a problem may sometimes be called for, such as minimizing the use of mutagenic, carcinogenic, or untested chemical pesticides, by education about potential hazards, product use, or alternatives, and by incentives, penalties, and taxes where necessary.

al., 1978; Wattenberg, 1978), promoters (Hiatt *et al.*, 1977; Miller *et al.*, 1979), and viruses (Hiatt *et al.*, 1977), which could have a great impact on prevention.

Mutagens and carcinogens must be identified and treated with respect*; priorities must be set, alternatives examined, and human exposure minimized. We have seen, and will continue to see, the folly of using humans as guinea pigs.

ACKNOWLEDGMENTS

This chapter first appeared as an article for *Science* (**204**:587–593, 1979). It is modified from a California Policy Seminar monograph prepared for the Institute of Governmental Studies, University of California at Berkeley, 1978.

This work has been supported by Department of Energy contract EY 76 S 03 0034 PA156. I am indebted to S. Kihara for help with the manuscript and to W. Havender, J. McCann, R. H. Harris, A. Blum, L. Gold, and N. K. Hooper for helpful criticisms.

REFERENCES

Ames, B. N., 1971, in: *Chemical Mutagens: Principles and Methods for Their Detection*, Vol. 1 (A. Hollaender, ed.), pp. 267–282, Plenum Press, New York.

Ames, B. N., and Hooper, K., *Nature (London)* **274**:19.

Ames, B. N., and Whitfield, H. J., Jr., 1966, *Cold Spring Harbor Symp, Quant. Biol.* **31**:221.

---

*The carcinogen and mutagen vinyl chloride is still used in the plastics industry to make vinyl floor tiles and PVC pipe, but vinyl chloride is no longer used in millions of cosmetic spray cans, and workers are no longer breathing in a dose that could give a high percentage of them cancer.

Ames, B. N., Gurney, E. G., Miller, J. A., and Bartsch, H., 1972a, *Proc. Natl. Acad. Sci. USA* **69**:3128.
Ames, B. N., Sims, P., and Grover, P. L., 1972b, *Science* **176**:47.
Ames, B. N., Kammen, H. O., and Yamasaki, E., 1975a, *Proc. Natl. Acad. Sci. USA* **72**:2423.
Ames, B. N., McCann, J., and Yamasaki, E., 1975b, *Mutat. Res.* **31**:347.
Benditt, E. P., 1977, *Sci. Am.* **236**:74.
Bjeldanes, L. F., and Chang, G. W., 1977, *Science* **197**:577.
Blum, A., and Ames, B. N., 1977, *Science* **195**:17.
Blum, A., Gold, M. D., Ames, B. N., Kenyon, C., Jones, F. R., Hett, E. A., Dougherty, R. C., Horning, E. C., Dzidic, I., Carroll, D. I., Stillwell, R. N., and Thenot, J.-P., 1978, *Science* **201**:1020.
Brown, J. P., Brown, R. J., and Roehm, G. W., 1977, in: *Progress in Genetic Toxicology* (D. Scott, B.A. Bridges, and F. H. Sobels, eds.), pp. 185–190, Elsevier/North-Holland, Amsterdam.
Bruce, W. R., Varghese, A. J., Furrer, R., and Land, P. C., 1977, in: *Origins of Human Cancer* (H. H. Hiatt, J. D. Watson, and J. A. Winsten, eds.), pp. 1641–1646, Cold Spring Harbor Laboratory, Cold Spring Harbor, N.Y.
Burnet, F. M. 1974, *Intrinsic Mutagenesis: A Genetic Approach to Aging,* Medical and Technical Publishing, Lancaster, England.
Cairns, J., 1975, *Sci. Am.* **233**:64.
Carter, L. J., 1976, *Science* **192**:240.
Commoner, B., Vithayathil, A. J., Dolara, P., Nair, S., Madyastha, P., and Cuca, G. C., 1978, *Science* **201**:913.
Crump, K. S., Hoel, D. G., Langley, C. H., and Peto, R., 1976, *Cancer Res.* **36**:2973.
Denda, A., Inui, S., and Konishi, Y., 1977, *Chem. Biol. Interact.* **19**:225.
Dillard, C. J., Litov, R. E., and Tappel, A. L., 1978, *Lipids* **13**:396.
Doll, R., 1977, *Nature (London)* **265**:589.
Doll, R., 1978, *Cancer Res.* **38**:3573.
Donahue, E. V., McCann, J., and Ames, B. N., 1978, *Cancer Res.* **38**:431.
Dougherty, R. C., and Piotrowska, K., 1976, *Proc. Natl. Acad. Sci. USA* **73**:1777.
Ehrenberg, L., Osterman-Golkar, S., Segerbäck, D., Svensson, K., and Calleman, C. J., 1977, *Mutat. Res.* **45**:175.
Fishbein, L., 1977, *Potential Industrial Carcinogens and Mutagens* (Publication No. 560/5–77–005), U.S. Environmental Protection Agency, Washington, D.C.
Gail, M., 1975, *J. Chronic Dis.* **28**:135.
Garner, R. C., Miller, E. C., and Miller, J. A., 1972, *Cancer Res.* **32**:2058.
Guess, H., Crump, K., and Peto, R., 1977, *Cancer Res.* **37**:3475.
Harris, S. G., and Highland, J. H., 1977, *Birthright Denied,* Environmental Defense Fund, Washington, D.C.
Hashimoto, Y., Shudo, K., and Okamoto, T., 1979, *Mutat. Res.* **66**:191.
Hiatt, H. H., Watson, J. D., and Winsten, J. A. (eds.), 1977, *Origins of Human Cancer,* Cold Spring Harbor Laboratory, Cold Spring Harbor, N.Y.

Hollstein, M., McCann, J., Angelosanto, F., and Nichols, W., 1979, *Mutat. Res.*, in press.
Hooper, N.K., Ames, B. N., Saleh, M. A., and Casida, J. E., 1979, *Science*, in press.
Inui, N., Nishi, Y., and Taketomi, M., 1978, *Mutat. Res.* **57**:69.
Jose, J. G., 1979, *Proc. Natl. Acad. Sci. USA* **76**:469.
Kapt, R. W., Jr., Picciano, D. J., and Jacobson, C. D., 1979, *Mutat. Res.* **64**:47.
Kier, L. D., Yamasaki, E., and Ames, B. N., 1974, *Proc. Natl. Acad. Sci. USA* **71**:4159.
Kleihues, P., and Bucheler, J., 1977, *Nature (London)* **269**:625.
Krotoszynski, B., Gabriel, G., O'Neill, H., Claudio, M. P. A., 1977, *J. Chromatogr. Sci.* **15**:239.
Lijinsky, W., and Taylor, H. W., 1977, in: *Origins of Human Cancer* (H. H. Hiatt, J. D. Watson, J. A. Winsten, eds.), pp. 1579–1590, Cold Spring Harbor Laboratory, Cold Spring Harbor, N.Y.
McCann, J., and Ames, B. N., 1976, *Proc. Natl. Acad. Sci. USA* **73**:950.
McCann, J., and Ames, B.N., 1977, in: *Origins of Human Cancer* (H. H. Hiatt, J. D. Watson, J. A. Winsten, eds.), pp. 1431–1450, Cold Spring Harbor Laboratory, Cold Spring Harbor, N.Y.
McCann, J., Choi, E., Yamasaki, E., and Ames, B. N., 1975, *Proc. Natl. Acad. Sci. USA* **72**:5135.
McMahon, R. E., Cline, J. C., and Thompson, C. Z., 1979, *Cancer Res.* **39**:682.
Maher, V. M., Miller, E. C., Miller, J. A., and Szybalski, W., 1968, *Mol. Pharmacol.* **4**:411.
Malling, H. V., 1969, *Genetics* **61**:s39.
Malling, H. V., 1971, *Mutat. Res.* **13**:425.
Marzulli, F. N., Green, S., and Maibach, H. I., 1978, *J. Environ. Pathol. Toxicol.* **1**:509.
Mau, G., and Netter, P., 1974, *Dtsch. Med. Wochenschr.* **99**:1113.
Maugh, T. H., II, 1978, *Science* **199**:162.
Mes, J., Campbell, D. S., Robinson, R. N., and Davies, D. J. A., 1977, *Bull. Environ. Contam. Toxicol.* **17**:196.
Meselson, M., and Russell, K., 1977, in: *Origins of Human Cancer* (H. H. Hiatt, J. D. Watson, and J. A. Winsten, eds.), pp. 1473–1481, Cold Spring Harbor Laboratory, Cold Spring Harbor, N.Y.
Miller, J. A., and Miller, E. C., 1977, in: *Origins of Human Cancer* (H. H. Hiatt, J. D. Watson, and J. A. Winsten, eds.), pp. 605–627, Cold Spring Harbor Laboratory, Cold Spring Harbor, N.Y.
Miller, J. A., Miller, E. C., Sugimura, T., Takayama, S., and Hirono, I. (eds.), 1979, *Naturally Occurring Carcinogens–Mutagens and Modulators of Carcinogenesis,* Proceedings of the 9th Princess Takamatsu Symposium, University Park Press, Baltimore, in press.
Mulvihill, J. J., Miller, R. W., and Fraumeni, J. F., Jr. (eds.) 1977, *Genetics of Human Cancer,* Raven Press, New York.

Nagao, M., Yahagi, T., Kawachi, T., Seino, Y., Honda, M., Matsukura, N., Sugimura, T., Wakabayashi, K., Tsuji, K., and Kosuge, T., 1977, in: *Progress in Genetic Toxicology* (D. Scott, B. A. Bridges, and F. H. Sobels, eds.), pp. 259–264, Elsevier/North-Holland, Amsterdam.

Nagao, M., Sugimura, T., and Matsushima, T., 1978, *Annu. Rev. Genet.* **12**:117.

Olson, W. A., Huberman, R. T., Weisburger, E. K., Ward, J. M., and Weisburger, J. H., 1973, *J. Natl. Cancer Inst.* **51**:1993.

Painter, R. B., and Howard, R., 1978, *Mutat. Res.* **54**:113.

Peto, R., 1979, *Proc. R. Soc. London Ser. B*, in press.

Prival, M. J., McCoy, E. C., Gutter, B., and Rosenkranz, H. S., 1977, *Science* **195**:76.

Purchase, I. F. H., Longstaff, E., Ashby, J., Styles, J. A., Anderson, D., Lefevre, P. A., and Westwood, F. R., 1976, *Nature (London)* **264**:624.

Purchase, I. F. H., Longstaff, E., Ashby, J., Styles, J. A., Anderson, D., Lefevre, P. A., and Westwood, F. R., 1978, *Br. J. Cancer* **37**:873.

Rannug, U., and Ramel, C., 1977, *J. Toxicol. Environ. Health* **2**:1019.

Rapoport, I. A., 1960, *Dokl. Akad. Nauk* SSSR **134**:1214.

Savage, E. P., 1976, *National Study to Determine Levels of Chlorinated Hydrocarbon Insecticides in Human Milk,* EPA Contract 68-01-3190, Colorado Epidemiological Pesticide Studies Center, Colorado State University, Fort Collins, Co.

Savage, E. P., 1977, *Polychlorinated Biphenyls in Human Milk,* Colorado Epidemiological Pesticide Studies Center, Colorado State University, Fort Collins, Co.

Segerbäck, D., Calleman, C. J., Ehrenberg, L., Löfroth, G., and Osterman-Golkar, S., 1978, *Mutat. Res.* **49**:71.

Shakarnis, V. F., 1969, *Genetika* **5**:89.

Shore, R. E., Pasternack, B. S., Thiessen, E. U., Sadow, M., Forbes, R., and Albert, R. E., 1979, *J. Natl. Cancer Inst.* **62**:277.

Sincock, A. M., 1977, in: *Origins of Human Cancer* (H. H. Hiatt, J. D. Watson, and J. A. Winsten, eds.), pp. 941–954, Cold Spring Harbor Laboratory, Cold Spring Harbor, N.Y.

Slater, E. E., Anderson, M. D., and Rosenkranz, H. S., 1971, *Cancer Res.* **31**:970.

Sugimura, T., Sato, S., Nagao, M., Yahagi, T., Matsushima, T., Seino, Y., Takeuchi, M., and Kawachi, T., 1976, in: *Fundamentals in Cancer Prevention* (P. N. Magee, S. Takayama, T. Sugimura, and T. Matsushima, eds.), pp. 191–215, University of Tokyo Press, Tokyo.

Tomatis, L., Agthe, C., Bartsch, H., Huff, J., Montesano, R., Saracci, R., Walker, E., and Wilbourn, J., 1978, *Cancer Res.* **38**:877.

Torkelson, T. R., Sadek, S. E., Rowe, V. K., Kodama, J. K., Anderson, H. H., Loquvam, G. S., and Hine, C. H., 1961, *Toxicol. Appl. Pharmacol.* **3**:545.

Van Duuren, B. L., Loewengart, G., Seidman, I., Smith, A. C., and Melchionne, S., 1978, *Cancer Res.* **38**:3236.

Viczian, M., 1969, *Z. Haut. Geschlechtskr.* **44**:183.
Wang, T., Kakizoe, T., Dion, P., Furrer, R., Varghese, A. J., and Bruce, W. R., 1978, *Nature (London)* **276**:280.
Wattenberg, L. W., 1978, *J. Natl. Cancer Inst.* **60**:11.
Wyrobek, A. J., and Bruce, W. R., 1975, *Proc. Natl. Acad. Sci. USA* **72**:4425.
Wyrobek, A. J., and Bruce, W. R., 1978, in: *Chemical Mutagens,* Vol. 5 (A. Hollaender and F. J. deSerres, eds.), Chap. 11, Plenum Press, New York.
Wyrobek, A. J., and Gledhill, B. L., 1979, in: *Proceedings,* Workshop on Methodology for Assessing Reproductive Hazards in the Workplace, National Institute of Occupational Safety and Health, Washington, D.C., in press.
Yamasaki, E., and Ames, B. N., 1977, *Proc. Natl. Acad. Sci. USA* **74**:3555.
Yoshida, D., Matsumoto, T., Yoshimura, R., and Matsuzaki, T., 1978, *Biochem. Biophys. Res. Commun.* **83**:915.

Closing Remarks

LEWIS THOMAS

I have to say a few things about the present state of medical science, and some possible implications for the future of the practice of medicine. This is risky ground. We do not have a very good record for the fulfilling of promises in medicine: over the past 25 years we have made too many promises, predicted too much too soon, and bragged much too much. As a result the public has mixed feelings about medicine as a scientific enterprise. On the one hand we are widely admired for having presumably converted medicine from an art to a science within a generation, but on the other we are confronted by uncomfortable questions about the human diseases which remain unsolved. How is it that such a high science, possessed (as it is generally believed) of a correspondingly high technology, can have left us still with such a list of things undone? More than half of human cancers remain beyond curing (although our methods for destroying cancer tissue have greatly improved); schizophrenia can be palliated by today's drugs, but it cannot be turned round or prevented; heart disease and stroke are still the major causes of premature

LEWIS THOMAS • President, Memorial Sloan–Kettering Cancer Center, New York, New York 10021.

death in our society; cirrhosis, rheumatoid arthritis, multiple sclerosis, the vascular lesions of diabetes, the senile psychoses, mental retardation, and the majority of congenital disorders all remain untouchable. With all that science, why such a list?

The truth is, of course, that medicine is still far from being a genuine, comprehensive science. It is now only at its earliest, most primitive beginnings: still pre-Newtonian, pre-Darwinian when compared with the physical and general biological sciences. We have nothing like a comprehensive understanding of the underlying mechanisms of human disease, even for the infectious diseases where we have come the closest to a technology based on genuine science. We are nowhere near the kind of information needed for unifying theory in medicine. We have nothing yet to match the astonishing surge of new knowledge achieved by the biological revolution in the past 25 years. Indeed, some of us in medicine have embraced the hybrid term "biomedical" science, partly in acknowledgment of the plain fact that our future depends on the surge in biology, but partly also, to be entirely candid, because of the protective coloring the term provides while we lag behind in the turbulence of biology's wake.

We cannot move in medicine until we have a fundamental understanding of disease, and we have no choice in this matter: we are compelled to wait until we have it, and information at this level comes in slowly, bit by bit. Without it, we are stuck. Empiricism has not really worked out successfully in medicine, although we have leaned on it for almost all the thousands of years of our professional existence. The most famous examples of empiricism at work—digitalis, quinine, opium, and ephedrine, for instance—are also the most extreme exceptions. Almost everything we used to do in the way of medical therapy was arrived in the course of trying this and that: an embarrassing amount of it was valueless and some of it downright harmful. Bleeding, cupping, purging, the use of every growable herb, the wholesale removal of teeth and tonsils for what were believed to be focal infections, even (in my view) the ritual incantations involved in psychoanalysis, were arrived at empirically—surely a mixed bag. We have a long record of trial and error, usually in precisely that order.

It is surprising that we have put up with all this without blushing for so long a time, and even more surprising that the public, our patients, have put up with it as well. Montaigne had no use at all for the medicine of his time but most people swallowed it without protest. It was not until one-third of the way through this century that we began to blush because of the sudden realization that there could be such a thing as science in medicine and that it could be useful. Vitamin B for pellagra, liver extract (later known to be vitamin B-12) for pernicious anemia, previously a 100% lethal disease, and insulin for diabetes were the earliest clues. Then, in the late 1930s came the sulfonamides and penicillin, and then streptomycin and all the other antibiotics—and we were off and running with science. Or so we believed, and so we began promising.

Medicine seemed to be transformed by these events. It is hard now to sense the magnitude of the change unless you were, as I was, a medical student during the time of the transition. In my day medical students were not trained by Harvard to treat human disease; there was a pervasive and powerful skepticism about all efforts at therapy. We were trained in the natural history of disease. The ultimate purpose of what was called "scientific medicine" in the mid-1930s was the making of an accurate diagnosis. This was the primary function of the physician, and having made the accurate diagnosis his work consisted of supporting and guiding his patients and their families while the disease, whatever, ran its natural course for better or worse. And there was no thought at all that things would change. Penicillin was greeted at first with disbelief, then astonishment. We had no idea such things could be.

Then, of course, we became converted, and ever since that time we and the public at large have been waiting, expectantly and impatiently, for comparable advances in other fields of medicine to match the revolutionary accomplishments in infectious disease. Some excellent things have happened, indeed: in endocrine-replacement therapy, in immunization against viral infection, in the prevention of hemolytic disease of the newborn, perhaps in the treatment of hypertension and Parkinsonism, in surgical technique, and a few others—but nothing like what we all wished for. And here we are, in 1979, and there is still atherosclerosis all around, still coronary occlusion, cancer, stroke, and the rest of that list. Why are these problems so difficult, considering the ease with which we gained so much control over microbial infection?

We tend to make up excuses for this asymmetry. We say that today's major illnesses are *chronic,* as though being chronic meant that the disease mechanism might be infinitely more subtle or obscure. Or we use terms like "multifactorial" or "environmental", with the implication that the problems are thereby made much less approachable, even insoluble. Or we pretend that they are simple problems, simply caused by wrong habits of living, and that by jogging or dieting or meditating or riding bicycles, and by abstention, the public could rid itself of these complex disabilities.

But the hard truth is that we are still too ignorant. We do not yet know enough to begin explaining the fundamental mechanisms of today's unsolved diseases, and we probably have a long way to go. The one thing we have learned in medicine, I hope, is that we will be required, absolutely, to have this basic information before we can undertake decisive, conclusive measures for treatment and prevention. This is, in fact, the lesson learned from the great advances in infectious disease.

Some of us have forgotten how this happened. We did not simply trip over penicillin or blunder our way into the capacity to eliminate tertiary syphilis and tuberculosis, or lobar pneumonia, or meningitis, or streptococcal disease. This was not empiricism at work. On the contrary, it was required that more than 60 years of the hardest kind of basic science be accomplished before the very notion of penicillin could make sense. Beginning around 1875, after it had been discovered that there were such things as bacteria and the even more flabbergasting discovery that certain bacteria caused certain entirely specific diseases, several generations of extraordinary scientists wore themselves out in the exploration of the details. Before their time, typhoid and typhus fever were thought to be aspects of the same illness, possibly related to malaria, caused by something wrong with the air. Tuberculosis and syphilis were believed to be a great number of different, totally obscure diseases, totally unapproachable. The childhood contagions were known to be contagious, but no one had the ghost of an idea what was being transmitted.

While the work of sorting out, defining, and classifying infectious agents was going on during all that 60 years, there was very little of what we would call "payoff" today. Immunization was recognized as a possibility, and therapeutic antisera were developed for diphtheria and certain pneumococcal infections, but that was about it. The investigators themselves must have known that their work would lead to something useful and practical someday, and some of them, like Robert Koch and Paul Ehrlich, were able to make a good theoretical case for this early on. But there was precious little proof throughout the 60 years, right up to the moment of prontosil and penicillin. It was a long, hard pull.

Something like this effort could still lie ahead of us in cancer or coronary disease or stroke. Or we could be as far from perceiving the underlying mechanisms as our pre-1875 forebears were in their perception of tuberculosis, and therefore we could still have ahead of us the equivalent of their six decades of research. Personally, I cannot believe this. I believe instead that we are already deep into it, far enough so that we should not be surprised by being surprised at any time from now on. We have already come some distance, as the papers just presented by my colleagues on this program have demonstrated. And we are, I firmly believe, on the right track.

I cannot think of any human disease that still has the look of impenetrability and unapproachability by scientific study. Indeed, for most of them the hard intellectual problem is deciding which of the several possible lines of approach is the best one to take because there are more choices than any of us would have predicted a few years back. The work is filled with uncertainties, wide open to surprise at every turn, which I take to be as good a way of evaluating basic research as any I can think of. Uncertainty is the element that defines basic research. When enough hard work has been done and there is enough certainty about the facts at hand, then and only then does applied research become possible. This is the way it has happened in the past in chemistry, physics, biology, even cosmology, and my guess is that it will be like this for medicine in the years ahead. This generation has the good luck to be involved in something close to the very beginning, in a field of science that is busy now with its own creation. If we can keep at it, taking advantage of new information as quickly as it comes into view, learning more whenever the opportunity arises, admitting ignorance more candidly than is the old habit of medicine, so that the public has a better feel for what we are up to and what the real stakes are, biomedical science will earn its keep among useful human endeavors.